Probability and Statistics (Continued)

MUIRHEAD • Aspects of Multivariate Statistical Theor
OLIVER and SMITH • Influence Diagrams, Belief N
*PARZEN • Modern Probability Theory and Its Applications
PRESS • Bayesian Statistics: Principles, Models, and Applications
PUKELSHEIM • Optimal Experimental Design
PURI and SEN • Nonparametric Methods in General Linear Models
PURI, VILAPLANA, and WERTZ • New Perspectives in Theoretical and Applied Statistics
RAO • Asymptotic Theory of Statistical Inference
RAO • Linear Statistical Inference and Its Applications, *Second Edition*
RAO and SHANBHAG • Choquet- Deny Type Functional Equations with Applications to Stochastic Models
RENCHER • Methods of Multivariate Analysis
ROBERTSON, WRIGHT, and DYKSTRA • Order Restricted Statistical Inference
ROGERS and WILLIAMS • Diffusions, Markov Processes and Martingales Volume I: Foundations, *Second Edition,* Volume II: Itô Calculus
ROHATGI • An Introduction to Probability Theory and Mathematical Statistics
ROSS • Stochastic Processes
RUBINSTEIN • Simulation and the Monte Carlo Method
RUBINSTEIN and SHAPIRO • Discrete Event System: Sensitivity Analysis and Stochastic Optimization by the Score Function Method
RUZSA and SZEKELY • Algebraic Probability Theory
SCHEFFE • The Analysis of Variance
SEBER • Linear Regression Analysis
SEBER • Multivariate Observations
SEBER and WILD • Nonlinear Regression
SERFLING • Approximation Theorems of Mathematical Statistics
SHORACK and WELLNER • Empirical Processes with Applications to Statistics
SMALL and McLEISH • Hilbert Space Methods in Probability and Statistical Inference
STAPLETON • Linear Statistical Models
STAUDTE and SHEARER • Robust Estimation and Testing
STOYANOV • Counterexamples in Probability
STYAN • The Collected Papers of T.W. Anderson: 1943-1985
TANAKA • Time Series Analysis: Nonstationary and Noninvertible Distribution Theory
THOMPSON and SEBER • Adaptive Sampling
WHITTAKER • Graphical Models in Applied Multivariate Statistics
YANG • The Construction Theory of Denumerable Markov Processes

Applied Probability and Statistics

ABRAHAM and LEDOLTER • Statistical Methods for Forecasting
AGRESTI • Analylis of Ordinal Categorical Data
AGRESTI • Categorical Data Analysis
AGRESTI • Introduction to Categorical Data Analysis
ANDERSON and LOYNES • The Teaching of Practical Statistics
ANDERSON, AUQUIER, HAUCK, OAKES, VANDAELE, and WEISBERG • Statistical Methods for Comparative Studies
*ARTHANARI and DODGE • Mathematical Programming in Statistics
ASMUSSEN • Applied Probability and Queues
*BAILEY • The Elements of Stochastic Processes with Applications to the Natural Sciences
BARNETT and LEWIS • Outliers in Statistical Data *Third Edition*
BARTHOLOMEW, FORBES, and McLEAN • Statistical Techniques for Manpower Planning, *Second Edition*

* Now available in a lower priced paperback edition in the Wiley Classics Linrary.

BATES and WATTS • Nonlinear Regression Analysis and Its Applications
BECHHOFER, SANTNER, and GOLDSMAN • Design and Analysis of Experiments for Statistical Selection, Screening, and Multiple Comparisons
BELSLEY • Conditioning Diagnostics: Collinearity and Weak Data in Regression
BELSLEY, KUII, and WELSCH • Regression Diagnostics: Identifying Influential Data and Sources of Collinearity
BERRY • Bayesian Analysis in Statistics and Econometrics: Essays in Honor of Arnold Zellner
BHAT • Elements of Applied Stochastic Processes, *Second Edition.*
BIIATTACHARYA and WAYMIRE • Stochastic Processes with Applications
BIEMER, GROVES, LYBERG, MATHIOWETZ, and SUDMAN • Measurement Errors in Surveys
BIRKES and DODGE • Alternative Methods of Regression
BLOOMFIELD • Fourier Analysis of Time Series: An Introduction
BOLLEN • Structural Equations with Latent Variables
BOULEAU • Numerical Methods for Stochastic Processes
BOX • R.A. Fisher, the Life of a Scientist
BOX and DRAPER • Empirical Model-Building and Response Surfaces
BOX and DRAPER • Evolutionary Operations: A Statistical Method for Process Improvememt
BOX, HUNTER, and HUNTER • Statistics for Experimenters: An Introduction to Design, Data Analysis, and Model Building
BROWN and HOLLANDER • Statistics: A Biomedical Introduction
BUCKLEW • Large Deviation Techniques in Decision, Simulation, and Estimation
BUNKE and BUNKE • Nonlinear Regression, Functional Relations and Robust Methods: Statistical Methods of Model Building
CHATTERJEE and HADI • Sensitivity Analysis In Linear Regression
CHATTEERJEE and PRICE • Regression Analysis by Example, *Second Edition*
CLARKE and DISNEY • Probability and Random Processes: A First Course with Applications, *Second Edition*
COCHRAN • Sampling Techniques, *Third Edition*
COCHRAN and COX • Experimental Designs, *Second Edition*
CONOVER • Practical Nonparametric Statistics, *Second Edition*
CONOVER and IMAN • Introduction to Modern Business Statistics
CORNELL • Experiments with Mixtures, Designs, Models, and the Analysis of Mixture Data, *Second Edition*
COX • A Handbook of Introductory Statistical Methods
*COX • Planning of Experiments
COX, BINDER CHINNAPPA, CHRISTIANSON, COLLEDGE, and KOTT • Business Survey Methods
CRESSIL • Statistics for Spatial Data, *Revised Edition*
DANIEL • Application of Statistics to Industrial Experimentation
DANIEL • Biostatistics: A Foundation for Analysis in the Health Sciences, *Sixth Edition*
DAVID • Order Statistics, *Second Edition*
*DEGTROOT, FIENBERG and KADANE • Statistics and the Law
*DEMING • Sample Design in Business Research
DILLON and GOLDSTEIN • Multivariate Analysis Methods and Application
DODGE and ROMIG • Sampling Inspection Tables, *Second Edition*
DOWDY and WEARDEN • Statistics for Research, *Second Edition*
DRAPER and SMITH • Applied Regression Analysis, *Second Edition*
DUNN and CLARK • Applied Statistics: Analysis of Variance and Regression, *Second Edition*

*Now available in a lower priced paperback edition in the Wiley Classics Library

Assessment: Problems, Developments and Statistical Issues

Assessment: Problems, Developments and Statistical Issues

A volume of expert contributions

Edited by

Harvey Goldstein
Institute of Education, London, UK

Toby Lewis
University of East Anglia, Norwich, UK

JOHN WILEY & SONS
Chichester · New York · Brisbane · Toronto · Singapore

Baffins Lane, Chichester,
West Sussex PO19 1UD, England

National 01243 779777
International (+44) 1243 779777

Other Wiley Editorial Offices

John Wiley & Sons, Inc., 605 Third Avenue,
New York, NY 10158-0012, USA

Jacaranda Wiley Ltd, 33 Park Road, Milton,
Queensland 4064, Australia

John Wiley & Sons (Canada) Ltd, 22 Worcester Road,
Rexdale, Ontario M9W 1L1, Canada

John Wiley & Sons (SEA) Pte Ltd, 2 Clementi Loop #02-01,
Jin Xing Distripark, Singapore 0512

Library of Congress Cataloging-in-Publication Data

Assessment : problems, developments, and statistical issues : a volume
of expert contributions / edited by Harvey Goldstein and Toby Lewis.
p. cm
Includes bibliographical references and index.
ISBN 0 471 95668 6 (alk. paper)
1. Educational tests and measurements. 2. Educational tests and
measurements—Statistics. 3. Examinations—Validity.
4. Examinations—Interpretation. 5. Test bias. I. Goldstein,
Harvey. II. Lewis, Toby.
LB3051.A767 1996
371.2′6—dc20

British Library Cataloguing in Publication Data

A catalogue record for this book is available from the British Library

ISBN 0 471 95668 6

Typeset in 10/12pt Times by Vision Typesetting, Manchester
Printed and bound in Great Britain by Biddles Ltd, Guildford and King's Lynn
This book is printed on acid-free paper responsibly manufactured from sustainable forestation,
for which at least two trees are planted for each one used for paper production.

Contents

Preface

Assessment, in all its forms, is in the news. It affects the lives and work of students, employees and institutions. Yet often it is poorly understood, its purposes confused and its design inadequate. In this volume we have brought together experts in a variety of areas of assessment to give the reader a picture of current issues, problems and developments with ample references to further work.

Not all our contributors share the same views, but that is as it should be. There exists no corpus of shared understandings and assumptions which would allow us to treat "assessment" as a discipline in its own right. Rather, it is a collection of separate strands, borrowing freely from other disciplines such as psychology, sociology and statistics, and it is this diversity which our contributions represent.

Our aim has been to make these contributions accessible to a wide audience and it is our hope that they will inform both academic and political debate.

Harvey Goldstein
Toby Lewis
October 1995

List of Contributors

Professor Patricia Broadfoot
School of Education
Bristol University
35 Berkeley Square
Bristol BS8 1JA
UK

Michael Cresswell
The Associated Examining Board
Stag Hill House
Guildford
Surrey GU2 5XJ
UK

Professor Harvey Goldstein
Institute of Education
20 Bedford Way
London WC1H 0AL
UK

Professor John Gray
Homerton College
Cambridge CB2 2PH
UK

Dr Vincent Greaney
Asia Technical Department
World Bank
Washington DC 20433
USA

Dr Thomas Kellaghan
Educational Research Centre
St Patrick's College
Dublin 9
Republic of Ireland

Professor Toby Lewis
Centre for Statistics
University of East Anglia
Norwich NR4 7TJ
UK

Professor George Madaus
Boston College
Campion Hall
Chestnut Hill
MA 02167
USA

Professor David B. McLay
Department of Physics
Queen's University
Kingston
Ontario K7L 3N6
Canada

Professor Leslie D. McLean
Ontario Institute for Studies in Education
252 Bloor Street West
Toronto
Ontario M5S 1V6
Canada

Anastasia Raczek
Boston College
Campion Hall
Chestnut Hill
MA 02167
USA

Dr Gillian Sutherland
Newnham College
Cambridge CB3 9DF
UK

Roderic Vincent
63 Woodfield Road
London W5 1SR
UK

Professor Alison Wolf
Institute of Education
London WC1H 0AL
UK

Geoffrey Woodhouse
Institute of Education
20 Bedford Way
London WC1H 0AL
UK

Biographies of Contributors

Patricia Broadfoot is Professor and Head of the School of Education and Director of the Centre for Curriculum and Assessment Studies at the University of Bristol. She is a specialist in comparative education as well as assessment, where she has published many books and articles. She has directed research projects on pupil profiling and records of achievement and is currently editor of the international journal *Assessment in Education.*

Correspondence: Professor P. Broadfoot, School of Education, Bristol University, 35 Berkeley Square, Bristol BS8 1JA, UK.

Michael Cresswell is Head of the Research and Statistics Group at the Associated Examining Board (England and Wales) and has been professionally involved in research related to educational assessment for over 20 years. His main current research project is concerned with the nature of examination standards, especially the qualitative judgements which underpin them. He is also involved in research into teachers judgements of their own students and has a long-standing interest in the problems of combining and equating assessments from instruments which differ in difficulty and type.

Correspondence: M. Cresswell, The Associated Examining Board, Stag Hill House, Guildford, Surrey GU2 5XJ, UK.

Harvey Goldstein is Professor of Statistical Methods at the Institute of Education, London. His two principal interests are in multilevel modelling of hierarchical data structures and in educational assessment. He has a particular interest in studies of school effectiveness and currently is directing a research project developing multilevel methodology and applications.

Correspondence: Professor H. Goldstein, Institute of Education, 20 Bedford Way, London WC1H 0AL, UK.

John Gray is Director of Research at Homerton College, Cambridge. Previously he was Professor of Education at Sheffield University. He has long-standing interests in research on school effectiveness and, more recently, on school improvement. His latest book (with Brian Wilcox) is entitled *Good School, Bad School; Evaluating Performance and Encouraging Improvement.*

Correspondence: J. Gray, Homerton College, Cambridge CB2 2PH, UK.

Vincent Greaney is a Senior Education Specialist in the World Bank. He has worked extensively in Ireland and in African and Asian countries. His areas of professional interest include public examinations, national assessment, reading development and teacher training.

Correspondence: Dr V. Greaney, Asia Technical Department, World Bank, Washington, DC 20433, USA.

Thomas Kellaghan is Director of the Educational Research Centre at St Patrick's College Dublin. He is currently Vice-president of the International Association for Educational Assessment.

Correspondence: Dr T. Kellaghan, Educational Research Centre, St Patrick's College, Dublin 9, Republic of Ireland.

Toby Lewis is Director of the Centre for Statistics at the University of East Anglia. He is a leading authority on statistical methods for "outlier" detection. He has a long-standing interest in assessment in higher education, especially matters to do with the moderation of examining procedures. He has also been involved in issues concerning evidence of misconduct and cheating in examinations.

Correspondence: Professor T. Lewis, Centre for Statistics, University of East Anglia, Norwich NR4 7TJ, UK.

George Madaus is the Boisi Professor of Education and Public Policy at Boston College. He is the former Director of the college's Center for the Study of Testing, Evaluation and Educational Policy. He has been President of the (US) National Council on Measurement in Education (NCME) and has served on US professional committees concerned with drafting test standards, and drafting codes of fair testing practices. He has been a visiting professor at the Harvard Graduate School of Education, a fellow of the Center for Advanced Study in the Behavioural Sciences and is a member of the National Academy of Education.

Correspondence: Professor G. Madaus, Boston College, Campion Hall, Chestnut Hill, MA 02167, USA.

David B. McLay was raised and educated in Hamilton, Ontario, with a master's degree in physics from McMaster University and a PhD from the University of British Columbia. His research field has been molecular spectroscopy and he has taught physics at the University of New Brunswick and Queen's University, Kingston, Ontario where he is now Emeritus Professor. For 10 years he represented Queens University on the Ontario Universities' Council on Admissions and was its chairman for two years.

Correspondence: Professor D. B. McLay, Department of Physics, Queens University, Kingston, Ontario K7L 3N6, Canada.

Leslie D. McLean is a Professor at the Ontario Institute for Studies in Education (OISE), in the Graduate Department of Education of the University of Toronto.

His research and development projects have included evaluation studies and surveys of student achievement, direction of Ontario's participation in the Second International Mathematics Study and research into mathematics/language relationships in curriculum and pedagogy.

Correspondence: Professor L. D. McLean, Ontario Institute for Studies in Education, 252 Bloor Street West, Toronto, Ontario M5S 1V6, Canada.

Anastasia Raczek is a doctoral student in educational research, measurement and evaluation at Boston College, and currently holds the Boisi fellowship in education and public policy.

Correspondence: A. Raczek, Boston College, Campion Hall, Chestnut Hill, MA 02167, USA.

Gillian Sutherland is Vice-Principal, Gwatkin Lecturer and Director of Studies in History at Newnham College, Cambridge. Much of her work hitherto has been concerned with the social history of education and is particularly interested in the experience of women and children. She is presently investigating changing attitudes to child and adolescent labour, as part of a larger project on constructions of childhood in nineteenth- and twentieth-century Britain. Her work on the history of assessment has led her to take a special interest in assessment in higher education and she is currently much involved in the analysis of the "gender deficit", the apparent under-performance of women in university examinations.

Correspondence: Dr G. Sutherland, Newnham College, Cambridge CB3 9DF, UK.

Roderic Vincent is Principal of Human Qualities and a partner in Tribute Publishing, working with large corporate clients in the areas of selection and assessment methods and human resources management. His work involves consulting in Europe and the Far East. He is a chartered occupational psychologist and an associate fellow of the British Psychological Society. He is a co-editor of *Selection and Development Review*, published by the British Psychological Society.

Correspondence: R. Vincent, 63 Woodfield Road, London W5 1SR, UK.

Alison Wolf is Professor of Education in the Department of Mathematical Sciences at the Institute of Education, London. Previously she worked for many years on education policy evaluations for the United States government and her current work is also closely involved with policy issues. She is an expert on vocational and professional assessment and qualifications, in a British and comparative context. Her current research combines quantitative studies with more general analyses of the incentives which assessment systems create for individuals and institutions.

Correspondence: Professor A. Wolf, Institute of Education, London WC1H 0AL, UK.

Geoffrey Woodhouse is a Lecturer in Statistics at the Institute of Education, London. He is interested in the methodology of multilevel modelling, especially as applied to educational assessment data. Currently he is investigating problems related to measurement errors in social data.

Correspondence: G. Woodhouse, Institute of Education, 20 Bedford Way, London WC1H 0AL, UK.

CHAPTER 1

The Scope of Assessment

Harvey Goldstein[1] and Toby Lewis[2]

[1]*Institute of Education, London, England*
[2]*University of East Anglia, Norwich, England*

INTRODUCTION

As Sutherland points out in Chapter 2 and Madaus and Raczek elaborate in Chapter 11, assessment has become a major industry during the twentieth century. Its history has been one of diverse and often quite separate developments. We have the psychometric tradition of highly technical test score modelling (Chapter 4), the carefully preserved system of curriculum-related public examinations described by Cresswell in Chapter 5, the diverse procedures used in workplace assessment discussed by Vincent in Chapter 15, and the often arcane procedures used in institutions of higher learning illustrated by McLay in Chapter 8. More recently there have been attempts to assess vocational aptitude, as Wolf discusses in Chapter 14, and attempts to involve individuals in their own assessment in order to make more authentic judgements, as described by Broadfoot in Chapter 3. Chapter 12 on assessment integrity, by Greaney and Kellaghan, leaves us in no doubt about the importance of assessment in the lives of individuals and this is closely related to another major concern of the late twentieth century, namely that of equity. Chapter 6 discusses these issues.

A recent development which has helped to raise the importance of assessment has been its increasing use for public accountability purposes. Chapter 7 looks at the notion of assessment "standards", and Gray in Chapter 9 and Woodhouse and Goldstein in Chapter 10 describe how schools are held to account on the basis of the assessment results of their students. Another aspect of the use of assessment for accountability lies at the international level, and McLean in Chapter 13 discusses the function of assessment surveys in comparing educational systems.

As the century nears its close the separate strands show signs of coalescing. This book is an attempt to present an overview of these different strands; to elucidate their similarities and differences.

Much of assessment is ostensibly quantitative, and even that which concentrates

Assessment: Problems, Developments and Statistical Issues.
Edited by H. Goldstein and T. Lewis.

on qualitative measurement necessarily encounters quantitative considerations when summarising or comparing what it finds. Much of the mathematical structure used is complex and Chapter 4 provides an overview of the major statistical procedures in the psychometric tradition.

In this introduction we do not set out to answer the question "what is assessment?" As will be clear from the range of the activities covered by our contributors, assessment is wide ranging and it is best defined by an enumeration of those activities themselves. To make sense of assessment it is illuminating to ask questions about its purpose, as do many of our authors. In the next section, therefore, we attempt to sketch a typology of purpose against which assessments can be judged.

PURPOSES OF ASSESSMENT

It is useful to distinguish three principal purposes or functions of assessment. The first is to certify or qualify individuals by discriminating among them; for example on the basis of a test, examination or teacher grading. The second is to assist in the learning process by providing an understanding of what someone has learnt so that remediation and further learning may take place. The third use is for making inferences about the functioning of institutions, enterprises or systems.

The certification or selection function is perhaps the oldest use of all. Its development was most rapid at the end of the nineteenth and beginning of the twentieth centuries when new ways for the mass selection of individuals were demanded. This function has remained perhaps the most important and also the most controversial of all, and the principal debates surrounding it have concerned the procedures which are most appropriate: few have advocated consistently that mass certification and selection are unnecessary. The debates about certification have focused on two major issues.

The first issue, which informed the early history, is that of comparability. As soon as certification is used on a large scale, for example for entry to higher education, some common "standard" is viewed as essential. At one extreme this has taken the form of highly formalised "standardised tests" which are entered under standard conditions and marked in a common way and where considerable efforts are devoted to the attempt to maintain "standards" across time and space. At the other extreme there has been a reliance on "professional judgement" to try to ensure an acceptable level of comparability, or on written guidelines which assessors are required to interpret and follow (Chapter 14). As the various contributions to this book make clear, all these efforts have problems associated with them. They all, in one way or another, rely upon particular assumptions about human behaviour and how observations of performance can be measured, classified and combined. In a quite general sense, the diversity of approaches reflects the cultural and political preferences of those who control assessment systems. In so far as this control is shared, for example among teachers, employers, governments and public servants, so there will emerge conflicts and compromises in what is effected. What is not in doubt is the requirement that those who promote or carry out certifications are required to justify their comparability and this is a theme which pervades all the writing on this topic and is explored by several contributors.

The role of assessment in learning has in one sense always existed since it is difficult to conceive of a learning process which is not concerned with monitoring its own progress. Broadfoot discusses in detail the subtle interplay here and reviews recent work and the tensions this has introduced. In particular she raises the issue of the potential conflict between this use of assessment and the first use.

The third function, that of comparing institutions or systems, has become extremely important most recently. There has been a long tradition of using assessment results comparatively to understand the reasons for institutional differences and even to try to explain differences between educational systems in different countries as McLean discusses in Chapter 13. This activity is perhaps best seen as one of research, with the measurement of relevant factors in an effort to explain observed variations. With the concern of governments to promote competition among educational institutions and with international comparisons becoming more focused on education and training, assessment results of all kinds are being used to judge performance. Yet as Gray argues, this is so fraught with difficulties that there are serious questions about its very legitimacy. Moreover, given the "high stakes" associated with such activity, there is a real danger that it may distort more legitimate and more worthy functions of assessment, and Broadfoot (Chapter 3) echoes some common concerns about this.

Evaluating assessment practices in terms of function can often help to structure the complexity of the surrounding issues. It can also help to avoid some of the confusion which may occur when functions are poorly conceptualised. An example of the failure properly to separate functions was in the use of so-called graded tests in the United Kingdom (Noss *et al.*, 1989) where the "learning promotion" function was deliberately linked to that of certification, with consequent devaluation of both.

Closely related to the notion of function is that of design. Thus, for example, for an assessment system whose aim is to provide comparative institutional information, there are strong constraints on the design, as discussed in Chapter 10. It needs, generally, to collect information on individuals across time with suitable measurements at the start and end of attendance at the institutions and can often be carried out using a sample of individuals. Certification procedures, on the other hand, do not require such longitudinal information, but potentially may apply to all individuals and will typically require further studies to assist in the search for comparability. Attempts to use, say, certification information to provide institutional comparisons for which they have not been designed directly leads to the serious problems discussed by Gray in Chapter 9.

ASSESSMENT AS A FUNCTION OF PURPOSE, METHOD AND THE PERSON ASSESSED

We have emphasised the different purposes of assessment, and the variety of different methods of assessment for achieving these purposes. But as well as purpose and method there is a third dimension to the structure—the person assessed. As Cresswell states in Chapter 5, "candidate attainment is itself affected by... characteristics of the candidates as individuals". Equally able individuals of different categories may well give different results on the same test.

What sort of categories are we talking about? What characteristics of candidates may affect the results of assessment? An obvious example is gender. Differences in performance between girls and boys are discussed in the book by Goldstein (Chapter 6), Cresswell (Chapter 5), Gray (Chapter 9). Another well-documented factor is the individual's ethnic group (see Chapter 6). Broadfoot (Chapter 3) brings out the distinction between personal factors such as ethnic background and gender, and factors residing in the social context such as school and classroom environment or again, as an indicator of standard of living, the taking or otherwise of free school meals.

The different characteristics of persons assessed may bear on their assessment in two ways. First, account may need to be taken of these individual factors in analysing or interpreting the results of a given assessment procedure. This problem is discussed at length in Chapter 6. Secondly, these factors may affect the choice or the design of an assessment procedure. From this point of view an extreme group, which we shall consider in some detail, is that of individuals with handicaps or special needs.

A candidate's test performance may be affected by adverse personal factors, ranging from "examination nerves" to major handicaps such as blindness. Assessment in such cases presents particular problems. How should it be carried out? How can we distinguish between special circumstances and features intrinsic to performance? In any substantial assessment system these problems cannot be left to *ad hoc* treatment of individual cases as they arise. This is well recognised by, for example, the City and Guilds of London Institute (C & G), one of the leading assessment and awarding bodies in the United Kingdom for work-related qualifications across all areas of industry and commerce (see Chapter 7). Their published *Policy and Practice* contains a "Special needs statement", which says:

> "City and Guilds takes a positive approach in providing a range of flexible and responsive assessment strategies and variations in methods of assessment. It aims to provide for candidates with disabilities and learning difficulties the same access to assessment as all other candidates. It seeks, as far as practicable, to remove barriers within the assessment which place candidates with special needs at a disadvantage, without thereby affording them an advantage over other candidates. In order to help achieve this aim, City and Guilds maintains contacts with professional bodies interested in both the occupational and educational opportunities for people with disabilities and learning difficulties" (C & G, 1994).

Table 1.1 Distribution of types of disability

Arthritis	629
Visual impairment	424
Hearing impairment	500
Cardiac conditions	311
Polio/spinal injury	222
Neurological disorders	356
Cerebral palsy	45
Asthma, bronchitis	146
Diabetes	162
Phobia/mental illness	546
Kidney/blood disorders	323
Epilepsy	171
Muscular dystrophy/atrophy	36
Stroke/brain disorders	168
Dyslexia	279
Disorders of muscles, joints	928
Other	328
Total: 4631 students with 5574 disabilities	

Consistently with the maintenance of their standards City and Guilds will, where possible, vary assessment arrangements for candidates presenting appropriate evidence of disabilities or learning difficulties. For example, arrangements for candidates with hearing impairment may include the use of a communicator/interpreter, extra time allowance, and mechanical/electronic aids. Candidates whose hearing loss results in a possible linguistic disability may be provided with question papers with appropriately modified wording, as recommended by a specialist teacher of the deaf. City and Guilds make corresponding arrangements for candidates with visual impairment, with medical conditions such as epilepsy, diabetes or respiratory disorders, and so on. In cases where a candidate is hospitalised or confined to home, arrangements can normally be made for an assessment to be held in hospital or at home.

The experience of the Open University (OU), the United Kingdom's nationwide distance learning institution, gives an idea of the extent of these problems. In 1994 some 4600 disabled students registered for OU courses, out of a total of some 102 800, i.e. 4.5% of registrations. The distribution of the various types of disability is reproduced here as Table 1.1, by kind permission of the OU (OU, 1994).

As with City and Guilds, the OU is prepared to make special arrangements for the assessment of students with handicaps. These include the use of an amanuensis or a typewriter/word processor and the provision of extra time or rest breaks. It is also possible to use examination papers in large print, on tape or in Braille, and to sit examinations at home or at a special centre, such as a hospital or institution. Naturally the OU stresses that, while there is flexibility in the arrangements that can be made, disabled students are subject to the same academic standards with regard to assessment as other students.

ASPECTS OF DESIGN

While there are different methods of assessment for different purposes, e.g. work sampling, ability tests, product appraisal and so on (see Chapters 7 and 15), we consider here the familiar situation of a certification assessment carried out by written test or examination. The previous section demonstrated the use of flexibility in the operation of a test when there were candidates with special needs; but of course there are always choices to be made in how to construct and apply a test irrespective of the individual characteristics of the candidates.

How can tests or examinations be designed for greatest efficiency? The word "design" here includes content, presentation, and conditions of administration. Some of the basic issues are discussed in Chapter 4. Here, without attempting to give answers, we list, by way of illustration, a few of the questions underlying efficient test and examination design.

1. What value is there, if any, in setting "bookwork" questions in examinations?
2. What are the uses of "open book" examinations? and of "take home" questions?
3. What choice of questions should be given within a paper?
4. What choice of papers should be given within a multi-paper examination?
5. What are the uses, if any, of essay-type questions in papers on science subjects?

All of these matters deserve serious attention. In the choice of questions, for example, practice varies widely among university departments, schools, etc. Typically a paper might contain (say) seven questions of which (say) five well answered would get full marks: one can ask whether candidates should be restricted to submitting not more than five answers. However, the policy of City and Guilds (see Chapter 7) is to "avoid offering choices of questions . . . since that militates against validity and reliability". To take another example, that of presentation, Chapter 4 points out that "apparently innocuous features of assessments, such as the layout of questions and certainly the language used in questions, can strongly influence responses".

The conditions of administration of a test or examination include the physical environment (e.g. the examination hall or other building, the seating, etc.) and the time of day. The choice of conditions may well affect the efficiency of the assessment, both absolutely and differentially between candidates with different characteristics (see Chapter 8).

IN CONCLUSION

In bringing together the contributors to this volume we have tried to do two things. First we have attempted to give the non-specialist reader an understanding of the activity known as assessment in different contexts. Secondly, we hope to convey an understanding that assessment serves many different purposes and many different people. Any assessment rests upon particular assumptions and exists within complex social structures and is informed by social and political preferences. In our view there are no "solutions" which can simultaneously satisfy competing requirements: assessment procedures are almost always compromises between barely reconcilable aims, whether these be authenticity vs. standardisation or single summaries vs. multidimensional description. The aim of this book is to promote a deeper understanding of the varieties of assessment and the multitude of contexts in which it has come to be used.

CHAPTER 2

Assessment: Some Historical Perspectives

Gillian Sutherland

Newnham College, Cambridge, England

INTRODUCTION

The historian of science, Stephen Jay Gould, remarked that "Science, since people must do it, is a socially embedded activity" (Gould, 1981, p. 21). At such a level of generality it sounds a banal comment, a truism; but it is one sharply relevant to a historical exploration of examinations and testing, "assessment" as it is currently called. For integral to many schemes of assessment have been techniques for which very aggressive claims have been made, that they are "scientific", for which read "objective", for which read "value-free". Historians, even those brought up within the British empiricist tradition, tend to be sceptical about such claims on principle and the history of assessment provides ample justification for such scepticism.

Assessment is a major twentieth-century industry. Its roots in the Western world go back to the late eighteenth century and are inextricably entangled both with the rise of formal schooling and with the development of the ideology of meritocracy. Any cluster of activities less likely to be value-free it would be difficult to imagine; although because these values continue to shape our society, their operation and potency may not always be easy to identify and measure.

It is a huge subject and an essay like this can only begin to scratch the surface. But in exploring some of the uses of assessment, mostly within the British Isles in the last two centuries, I hope to expose some of these entanglements and, in so doing, illustrate two propositions. First, tests and examinations have been not ends in themselves but means to ends; these ends have varied enormously. Second, when tests and examinations have been operated on a large scale, on large populations with many practitioners, both the practice and the end results have tended to be rather different from those initially envisaged by the planners. These are propositions which have a relevance beyond the parochial examples from which they have been developed and are ones which contemporary

Assessment: Problems, Developments and Statistical Issues.
Edited by H. Goldstein and T. Lewis.

practitioners of assessment may find illuminating. In offering them I have no wish to imply that contemporary dilemmas and situations are simply iterations of historical ones; rather to urge that the complexities, contingencies and sheer untidiness of the historical examples encourage the contemporary analyst to ask some more and some different questions.

Three uses of assessment have dominated its history in Britain so far: as a device for raising standards in schools and universities; as a device for measuring deviation or abnormality; as a device for securing equitable treatment. Conceptually and practically they have been linked, the first and the last most frequently yoked together; but for heuristic purposes it seems worth disentangling them.

RAISING STANDARDS

The use of assessment to raise and/or ensure standards is the easiest thread to unravel first. It was a key objective of the formal written examinations which gradually came to shape education in English universities in the nineteenth century. *Viva voce* or oral examinations have had a long pedigree, stretching back to medieval disputations. In Cambridge these developed into the university's Senate House Examination, by the mid-eighteenth century a written test in mathematical reasoning. In 1765 St John's College, concerned about the performance of its own students, instituted regular twice-yearly domestic tests for them all. In 1800 the University of Oxford enacted a statute providing for the examination of candidates for honours degrees. By 1825 the numbers of these candidates were sufficiently great for the examiners to require the answers to their questions in writing; and in 1828, for the first time, the question papers were printed (Roach, 1971, pp. 13–16; Carr-Saunders and Wilson, 1933, p. 311). In 1836 the University of London was created by a royal charter, as an examining body to secure uniform and appropriate standards in fledgeling institutions of higher education in the United Kingdom and in the Empire (Harte, 1986, pp. 23–61). The thrust of all these activities was well conveyed by the Royal Commissioners investigating the affairs of the University of Oxford in 1852, who commented, "to render a system of examination effectual it is indispensable that there should be danger of rejection for inferior candidates, honourable distinctions and substantial rewards for the able and diligent" (Carr-Saunders and Wilson, 1933, p. 309).

As the numbers aspiring to a higher education were increasing, so were those seeking —and attempting to provide —formal schooling at less exalted levels. The late eighteenth and early nineteenth century saw extensive attempts to reshape and reactivate ancient charitable trusts for the provision of grammar schools as well as the emergence, with a great surge of activity in the 1840s, of a new type of school, the proprietary boarding school, sometimes the speculative venture of one individual, but more often supported by a joint stock company (Sutherland, 1990, pp. 132–7). With the expansion of demand and provision came anxiety about the quality of that provision, issuing in the creation in 1857 of the Oxford Local Examinations Board and its Cambridge counterpart, the Local Examinations Syndicate. They offered written examinations for which either individual pupils or whole forms of schools might be entered. By the end of the 1860s, those campaigning for the improvement of educational provision for girls had succeeded in securing their admission to the Locals on the same terms as boys. In 1871 the Prize Day Report of the North London Collegiate School, whose headmistress was the formidable Frances Mary Buss, noted that "the level of scholarship in the higher classes has been steadily rising since the University of Cambridge extended its examinations to girls" (Roach, 1971, pp. 77–135).

Parallel with increasing demand for and concern about secondary education, the education of the middle classes as the nineteenth century defined it, there was a growing demand and concern for elementary education, the education of the working classes. The government first offered grants in support of elementary education in 1833. Demand for these, and consequently government expenditure, rose steadily, but there were claims and counter-claims about what was happening in the schools themselves: some alleged that teachers spent all their time on the brighter children, neglecting the average and weaker ones. Was greater expenditure being matched by more and/or better education? Were more and better the same thing? In 1862 Robert Lowe, the responsible minister, proposed a new system of grants for elementary schools, the bulk of which would depend on the performance of each child each year in examinations in reading, writing and arithmetic, conducted by the government inspectorate. The examination "standards" were age-related, and no child was to be allowed to sit the examinations of the same standard twice. The inspectors, who hitherto had regarded themselves primarily as missionaries of good practice, were horrified and there was a chorus of opposition. Lowe stood his ground, however, and payment by results, as it came to be known, prevailed for the next 30 years (Sutherland, 1973, pp. 4–9).

Lowe rejected the allegation that his covert objective was to cut expenditure. He insisted that his intention was to raise standards, by which he meant that teachers should be driven by threat of financial penalty to bring the largest possible number of children up to a minimum level of attainment. He defended his Revised Code of grant regulations as applying the principles of the free market to schooling:

> "What is the object of inspection? Is it simply to make things pleasant, give the schools as much as can be got out of the public purse, independent of their efficiency; or do you mean that our grants should not only be aids, subsidies, and gifts, but fruitful of good? That is the question, and it meets us at every turn. Are you for efficiency or for a subsidy? Is a school to be relieved because it is bad and therefore poor, or because it is a good school, and therefore efficient and in good circumstances? (*Hansard's Parliamentary Debates*, 13 February 1862, 205)

The immediate effect of the Revised Code was sharply to curb government expenditure on elementary schools. It fell for two years (1863–65) and did not again reach the levels of 1861 until 1869. Whether standards overall "rose" and/or teachers divided their time more evenly between their pupils it is impossible to judge. The formal class size was huge; problems in securing regular attendance meant that actual class size fluctuated wildly. In addition the only source of evidence we have for monitoring both "standards" and the behaviour of teachers are the reports of the government inspectors. They had varied in their views of these before 1862 and continued to do so after. From 1858 onwards inspectors' reports became increasingly subject to censorship. "Before" and "after" comparisons were vitiated by the fact that payment by results radically altered the relationship of the inspectors to the schools. Before, good inspectors had been guides and mentors; after, whether they liked it or not, they took on some of the characteristics of inquisitors. As one headmaster put it to a Royal Commission in 1887, the inspector's main duty was "to criticize and fine" (Cross Minutes, PP 1887, xxix, q. 23648; Sutherland, 1973, pp. 55–75).

Other consequences of these mass annual examinations are easier to establish. The relationship of the teacher to pupils, school authorities and government was also altered. Many school managers made a fraction of teachers' salaries dependent upon their "earnings" for their schools in the annual examination, driving them to draw the analogy with factory workers on piece rates. Many teachers "taught to the test", making each day almost a rehearsal for examination day. When the system had finally been dismantled, a former Senior Chief Inspector, Edmond Holmes, looked back:

> "When inspectors ceased to examine (in the stricter sense of the word) they realised what infinite mischief the yearly examination had done. The children, the majority of whom were examined in reading and dictation out of their own reading-books (two or three in number as the case might be), were drilled in the contents of those books until they knew them almost by heart. In arithmetic they worked abstract sums, in obedience to formal rules, day after day, and month after month; and they were put up to various tricks and dodges which would, it was hoped, enable them to know by what precise rules the various questions on the arithmetic cards were to be answered. They learned a few lines of poetry by heart, and committed all the 'meanings and allusions' to memory, with the probable result—so sickening must the process have been—that they hated poetry for the rest of their lives. In geography, history and grammar they were the victims of unintelligent oral cram, which they were compelled, under pains and penalties, to take in and retain till the examination day was over, their ability to disgorge it on occasion being periodically tested by the teacher. And so with the other subjects. Not a thought was given, except in a small minority of the schools, to the real training of the child, to the fostering of his mental (and other) growth. To get him through the yearly examination by hook or by crook was the one concern of the teacher. As profound distrust of the teacher was the basis of the policy of the Department, so profound distrust of the child was the basis of the policy of the teacher. To leave the child to find out anything for himself, to work out anything for himself, to think out anything for himself, would have been regarded as a proof of incapacity, not to say insanity, on the part of the teacher, and would have led to results which, from the 'percentage' point of view, would probably have been disastrous" (Holmes, 1911, pp. 107–8).

Ironically, the factors which brought about the end of payment by results had little to do with a principled critique of this kind and much more to do with administrative and political imperatives. It was dismantled less because it was perceived to be educationally stunting, than because it had become desperately unwieldy, bidding fair to collapse under its own weight.

Attempts to render the system less narrow from 1867 on had led to the addition of a host of other little grants for further subjects. This compounded the basic problem: a system which after the Education Act of 1870 had become a national system of grant, not simply one in aid of local pockets of voluntary endeavour, was dependent upon and determined by literally millions of individual encounters between the corps of inspectors and the children in the elementary school population—over 3½ million by 1890. From the end of the 1870s the Treasury and the Education Department were in a state of escalating warfare over the cost of the manpower needed to run such a system. In 1878 the Exchequer and Audit Department and the Public Accounts Committee fired the first shots in a similarly escalating war about the impossibility of monitoring and auditing all these separate payments. Lowe had imagined an elegantly self-balancing mechanism. The Treasury and Exchequer and Audit Department saw an uncontrollable bureaucracy with endless opportunities for confusion, misrepresentation and even petty pilfering (Sutherland, 1973, pp. 192–260). Meanwhile, at the beginning of the 1890s discussion was under way about the possibility of state funding for secondary education. (Sutherland, 1990, p. 151) The existing system of grants was hardly one on to which this could be grafted: no politician of sense would wittingly extend a battleground.

MEASURING ABNORMALITY

Payment by results was preoccupied with national minima in terms of attainments. From 1880 attendance at elementary school had been compulsory for the children of the working class in England and Wales; and this legislation gradually brought into schools children who appeared not to be able to cope with its demands at all, handicapped as they were by physical or mental disability (see Chapter 1). Their presence and the need to provide constructively for them fed a preoccupation with the diagnostic assessment of abilities, an attempt to measure and to express how far an individual's abilities deviated from the norm. The earliest attempts at this had been very crude, making much of external physical features, the "stigmata" of handicap and entailing elaborate tests of sensory perception (Sutherland, 1984, pp. 5–53).

Breakthrough to something far more sophisticated came only at the turn of the century, in France, when Alfred Binet and his collaborator Victor Simon, dissatisfied with tests which were both crude and indirect, set out to devise a test battery which directly engaged the higher mental processes. These tests were tried out, revised and standardised on hundreds of Parisian schoolchildren. They were graduated by age and Binet developed the concept of "mental age" which could be set beside the child's chronological age to provide a way of expressing how far the child's abilities deviated from the norm (Wolf, 1972, Ch. 4 and 5).

A child's abilities could, of course, far exceed the norm as well as fall below it. In addition, the range and variety of the tests, spatial, numerical, verbal—the mere use of a battery—offered some possibility that particular strengths and/or weaknesses could be identified, that the bald statement of mental age could be qualified by some sort of profile. The mode of administration recommended for these tests, published between 1905 and 1908, strengthened this possibility. Binet and Simon designed the tests to be administered one-to-one, creating a framework for dialogue within which a skilled and experienced tester could develop a relationship with a child.

Diagnostic testing of this kind was—and is—labour-intensive and expensive. Although the English Education Department's Chief Medical Officer took up Binet and Simon's tests with alacrity, response among his colleagues was patchy (Sutherland, 1984, pp. 57–85). A development which attracted more interest and attention in the Anglo-Saxon world was the experiments in England by W. H. Winch and Cyril Burt, and in California by Lewis Terman, to build group tests on the Binet–Simon model. These were written, pencil and paper tests of ability which could be administered *en masse*, the answers, right or wrong, then being scored by unskilled testers equipped with a key, or even by a machine. Such a test was a very blunt instrument in comparison with the individually administered Binet–Simon test, intended to draw, with as much precision as possible, the profile of an individual. But it had considerable attractions for those wanting to sort and classify large populations.

A perfect opportunity to demonstrate the use of group tests in this way presented itself in 1917, when the United States entered the First World War. The newly emerging professional group, the psychologists, made their contribution to the war effort by devising a set of tests, the Alpha and Beta tests, intended to help the Army distinguish speedily between those recruits who could not be trusted with a rifle and those who might be officer material.

A power struggle within the emergent profession of psychology played a major role in the formulation of the Alpha and Beta tests (Kevles, 1968), signalling the importance of vested interests in shaping the fortunes of any large-scale testing operation (see Chapter 11). In the first substantial commissions for the development of group tests in the UK, more apparently disinterested motives also played a part.

At the beginning of the 1920s two Local Education Authorities (LEAs) in England, faced with the necessity of selecting children from elementary schools (or primary schools as they were beginning to be called) to go on to secondary schooling, wished to do so as "fairly" as possible. They were worried about the ways in which existing tests of attainments appeared to favour children from more affluent backgrounds and were attracted by the claims being made for the new group tests that they were measuring ability rather than schooling. In 1919 the Education Committee of the Labour-controlled County Borough of Bradford invited Cyril Burt, then working for the London County Council, to devise a group test for use in their secondary school selection procedures. In 1920

Northumberland Education Committee, worrying in particular about the poor showing of rural children in their examination, invited Godfrey Thomson, Professor of Education at Newcastle, to do likewise for them. Thomson put the fees paid him by Northumberland, and other LEAs which copied them, into a trust fund to finance the development and publication of further tests, not only group "mental" or "intelligence" tests as they rapidly became labelled, but also standardised tests of English and Arithmetic. After his move to the Chair of Education and the Principalship of Moray House College in Edinburgh in 1925, these became familiar as the Moray House Tests (Sutherland, 1984, pp. 133–40).

So began in the UK the rise of the intelligence test and its use in the process of secondary school selection. So, too, the invention of the group test linked the methods and techniques of the psychology of individual differences into the general discourse on examinations. In so doing, it brought powerful reinforcements to those minded to argue for examinations as tools of equity.

SECURING EQUITY

The roots of the argument for examinations as tools of equity go as far back as those of the argument for examinations as a means of raising standards and they were frequently linked. The link is plain in the above-quoted comment of the Oxford Royal Commission of 1852: "to render a system of examinations effectual it is indispensable that there should be danger of rejection for inferior candidates, honourable distinctions and substantial rewards for the able and diligent". The examination could not only establish minimum standards; conducted competitively, it could also offer a means of identifying and rewarding talent. The reward of talent was perceived as particularly desirable because talent was equated with virtue, ability with merit.

One of the earliest but also one of the fullest and most resonant statements of this equation came from the historian Macaulay, speaking in the House of Commons in 1833, in support of the first attempt to introduce examinations into the processes of selection for the Indian Civil Service:

> "It is proposed that for every vacancy in the civil service four candidates shall be named, and the best candidate selected by examination. We conceive that under this system, the persons sent out [to India] will be young men above par, young men superior either in talents or in diligence to the mass. It is said, I know, that examinations in Latin, in Greek and mathematics, are no tests of what men will prove to be in life. I am perfectly aware that they are not infallible tests; but that they are tests I confidently maintain. Look at every walk of life and see whether it be not true that those who attain high distinction in the world were generally men who were distinguished in their academic career.... Education would be mere useless torture, if, at two or three and twenty, a man who had neglected his studies were exactly on a par with a man who had applied himself to them, exactly as likely to perform all the offices of public life with credit to himself and with advantage to society. Whether the English system of education be good or bad is not now the question. Perhaps I may think that too much time is given to the ancient languages and to the abstract

sciences. But what then? Whatever be the languages, whatever the sciences, which it is, in any age or country the fashion to teach, the persons who will become the greatest proficients in those languages and those sciences will generally be the flower of the youth, the most acute, the most industrious, the most ambitious of honourable distinctions" (Macaulay, 1898, xi, pp. 571–3).

Viewed thus, examinations were the obvious method of attacking patronage, hitherto the dominant mode of recruitment to all forms of government service and to most occupations, now increasingly being seen as corruption incarnate. Macaulay lost the battle in 1833, succeeding only 20 years later, with the India Act of 1853. But in the same year a committee consisting of Macaulay's brother-in-law, Sir Charles Trevelyan, and Sir Stafford Northcote, were preparing to recommend the adoption of the same principle in appointments to the Home Civil Service. Welcoming their report in 1854, the Chancellor of the Exchequer, W. E. Gladstone, prize and scholarship winner at Eton and Oxford, with Double First Class Honours in Classics and Mathematics, presented the abolition of patronage and the use of examinations for entry to government service as "a signal proof of cordial desire that the processes by which government is carried on should . . . be honourable and pure . . . a striking sign of national confidence in the intelligence and character of the people" (Morley, 1903, i, p. 610).

Competitive examination was seen not only as the instrument to clean up government service but also as the means of articulating the different parts of the educational system into a whole. In 1871 the scientist Thomas Henry Huxley told his fellow members of the London School Board, responsible for elementary education in the capital, that he

"should like to have an arrangement considered by which a passage could be secured for children of superior ability to schools in which they could obtain a higher instruction than in the ordinary ones. He believed that no educational system in the country would be worthy the name of a national system, or fulfil the great objects of education, unless it was one which established a great educational ladder, the bottom of which should be in the gutter and the top in the University and by which every child who had the strength to climb might, by using that strength, reach the place intended for him" (Philpott, 1904, pp. 153–4).

The class-based nature of provision, however, made the actual construction of a "scholarship ladder", as it became labelled, a slow process. Elementary and secondary schools operated as separate, parallel tracks and switching points were very rare. By 1900 only 5500 scholarships existed to allow bright children from the elementary track to jump across to the secondary track, whether represented by grammar schools supported by charitable endowments or private boarding schools. The creation of maintained, that is, government-funded secondary schools under the provisions of the Education Act of 1902 began to change this. Now there were elementary and secondary schools being run by the same local authority. The government's Free Place Regulations of 1907 gave a further boost, in providing extra grant for those secondary schools prepared to take 25% of their pupils from elementary schools without payment of fees. By 1912 just over 49 000 children, 32% of the population of maintained secondary schools, were ex-elementary school pupils holding scholarships or free places (Sutherland, 1984, pp. 108–11).

Those Free Place Regulations created the single most important rung of the scholarship ladder. The officials who framed them had envisaged that schools accepting the enhanced grant would set a simple qualifying examination for their "free-placers". Such was the demand, however, that this turned almost overnight into a fiercely competitive enterprise, usually taking place when the children were aged between 11 and 12. This was the origin of the "11-plus", the examination for entry to a selective secondary school, which dominated English maintained education for the first two-thirds of the twentieth century, the examination to which intelligence tests began to be added from the beginning of the 1920s.

Between 1919 and 1939 between a half and three-quarters of the LEAs in England and Wales with responsibility for secondary education used at some stage something they called an intelligence test; and use continued to expand through the war years, up to 1950. To say that, however, is not to say that by 1950 the majority of LEAs in England and Wales were committed to meritocracy. The actuality was a great deal less tidy and coherent than any quantitative generalisations about test use might suggest. Exploration of exactly what individual LEAs did when they said they were using tests is to uncover extraordinary variety, idiosyncrasy, often downright muddle and sometimes practices in head-on conflict with the ideals to which they were paying lip-service (Sutherland, 1984, pp. 164–269).

If any generalisations about the variety of practice in the inter-war period are possible at all, they are that in the 1920s use of the term "intelligence test" represented only the invocation of a fashionable catch-phrase; actual practice changed little. Sometimes the use of the catch-phrase appears designed to cloak the absence of any substantive change. In the 1930s some sophistication in examining techniques in general began slowly to spread, an impression paradoxically strengthened by an investigation of some of the LEAs (Hull, Sheffield and the giant pacemaker, the London County Council), where the decision not to use an intelligence test had, after careful consideration, been taken. There, it was argued that its addition would not sufficiently improve the success rate achieved through sophisticated attainment testing. As Dr A. G. Hughes, one of the officers most closely involved in London, commented, "No method which is reasonably simple administratively, can do more than rough justice to the candidates" (Sutherland, 1984, p. 262).

Behind this lay not only a sharp awareness of the practicalities of testing a large population but also a certain scepticism about the more extravagant claims made for intelligence tests to be "culture-free", to reveal something that sophisticated attainments tests did not reveal. The best of the specialist practitioners, Thomson included, had been aware of this from the beginning. But the overlap between the findings of intelligence and attainments tests was seldom complete; and they reasoned that, if selection was to be the order of the day, then it should be carried out with the most sophisticated battery of tests available. Godfrey Thomson himself defined the issue as "how with most justice to select eleven-year-old children in the primary school for the privilege of free secondary education. It was a problem which had a personal interest for me, for ... I would myself have had no education beyond the primary school had I not won a free place in a secondary school in a competitive examination" (Thomson, 1952, pp. 284–5).

Wartime and the immediate post-war period brought a new resonance to the discourse of meritocracy. The shared suffering of wartime and the sense of new beginnings at the war's end made "democratization" and "equality of opportunity" key phrases in the reform of the education system. The 1944 Education Act was proclaimed as bringing about "secondary education for all". The school-leaving age was raised and a major school building programme was launched. This expanded secondary education was, however, to be delivered in different kinds of schools; and the immediate effect of the post-war expansion was to bring a new uniformity and a new sophistication to 11-plus examinations. Almost all now entailed standardised English, arithmetic and intelligence tests and they were supposed to allocate children to grammar, technical and modern schools. Formally, these schools enjoyed "parity of esteem"; in actuality the hierarchy was plain and the examination functioned to select for the grammar school as before (Thom, 1986; Saint, 1987; Hennessy, 1992, pp. 155–62).

At the beginning of the 1950s meritocracy, on the face of it, appeared more attainable. Yet it was also becoming increasingly plain that measured ability, the central component of merit, was being found in ample supply among the middle and upper classes but less commonly among the working classes. In places as widely different as Middlesbrough and south-west Hertfordshire in 1953, only 9% of the sons of unskilled manual workers among the 10–11 age-group were offered grammar school places (Floud *et al.*, 1956, p. 42; Sanderson, 1987, pp. 45–7). Skewness on this scale made claims about equality of opportunity difficult to sustain.

Simultaneously the point was being made that the scholarship ladder did not take the scholarship holder clear of the class structure: it allowed him, or more rarely her, to move within it. In his brilliant satirical essay, *The Rise of the Meritocracy 1870–2033*, which actually coined the term, Michael Young contended that the role model for the successful meritocrat remained the "gentleman of independent means" (Young, 1958, p. 28). Richard Hoggart spelled out how, for the working-class scholarship child, this meant learning to behave like the middle class. He wrote poignantly of the working-class scholarship boy as being "at the friction-point of two cultures", "between two worlds of school and home; and they meet at few points. Once at the grammar school, he quickly learns to make use of a pair of different accents, perhaps even two different apparent characters and differing standards of value" (Hoggart, 1957, pp. 239, 242). Far from dissolving or at least diluting traditional structures of class, the ideology of meritocracy and its central mechanism, the scholarship ladder, appeared to be enabling the middle classes to engage most effectively in cultural reproduction. (Karabel and Halsey, 1977; Halsey *et al.*, 1980, Ch. 5).

Such realisations, and awareness that the 11-plus was still a selective device very much as it had been from its inception, fuelled a challenge to selection and its associated apparatus of tests far more comprehensive than anything that had gone before. The idea of a common secondary school had been first canvassed in Britain in the 1930s. By the beginning of the 1960s it began to look like practical politics; and from 1965 LEAs were encouraged to reshape their provision accordingly (Simon, 1991, esp. pp. 203–20, 271–301).

In the dock, along with selection, was the associated apparatus of testing. Much of the public debate drew no distinctions between formative and summative uses of assessment or perceived that it might have professional as well as managerial functions. The net effect was substantially to weaken the association, hitherto close and strong, between the technology of assessment and the language of equity. In some quarters, "testing" acquired heavily pejorative overtones (Simon, 1978; Boyle *et al.*, 1971, p. 23; Booth, 1983, p. 255). From the late 1970s assessment came instead to be more strongly associated in the public mind with a new phase of debate about "standards" and "accountability", deserving of a separate essay in itself (Simon, 1991, Pt. III; Gipps and Goldstein, 1983; Gipps *et al.*, 1983; Lawton and Gordon, 1993, pp. 12–19, 23–7).

As some state-funded secondary schools move back to selection at 11-plus in the 1990s, it will be interesting to see how far the association between the technology of assessment and the language of equity is revived. Perhaps the appeal by the High Master of Manchester Grammar School, now an independent school, for state funding for entrance scholarships to his school is a straw in the wind: "we support selection on academic merit" he declares, "as strongly as we oppose selection on social or economic grounds" (*Guardian*, 29 March 1995, p. 22). Would that such distinctions could be clearly and confidently drawn! The experience of the last 50 years in England points to some of the difficulties.

The practice of assessment is "socially embedded": people do it to other people —or at least, to other people's children. We shall only understand it fully if we take account of the social, economic and political contexts in which it grows and is practised. The destruction of payment by results owed much more to political and bureaucratic pressures than to ideology. We shall not understand the English version of meritocracy unless we grasp the continuing resilience of the English class structure, the peculiar legacy of being the first industrial nation and the decentralisation until the 1980s of the government of education in England. Tests and examinations are means, not ends. When operated on large populations, with many practitioners, they also exhibit other characteristics in common with large enterprises of social engineering: both the practice and the end results tend to turn out to be rather different from those envisaged by the planners.

This essay has attempted to illustrate these general propositions with a cluster of examples, drawn mostly from Britain and explored in some detail in a deliberate effort to show something of the complex interactions of process with context. The propositions themselves have more than parochial relevance and could, if space allowed, be pursued in other societies and other times. It does not seem inappropriate to suggest, for example, that the campaign for new standards in American schools (Resnick and Resnick, 1992) and the range of state responses to it derives some of its impetus from the slowing of American economic growth and associated cultural anxieties of a very deep-rooted kind. In France the increasing rigidity of streaming in the secondary sector from the end of the 1960s has combined with a deeply conservative pedagogy to facilitate the subversion of the democratisation which the *collège unique*, the common middle school, was intended to bring (Prost 1992, pp. 88–95). Those currently engaged in implementing successive versions of the National Curriculum and attempting its assessment in England, Wales, Scotland and Northern Ireland find themselves actors in a particularly large-scale enterprise of social engineering. The technology of assessment cannot be considered in isolation from the values of the educational system and the society in which it is intended to operate.

CHAPTER 3

Assessment and Learning: Power or Partnership?

Patricia Broadfoot

Bristol University, Bristol, England

ASSESSMENT FOR LEARNING: A TEACHER–STUDENT PARTNERSHIP

Assessment in education has many potential purposes. It is an integral part of the teaching–learning process providing feedback for both teachers and students which can guide decisions concerning future learning goals. In addition to its central role in the *curriculum*, assessment also has a key role to play in *communicating* information about students' achievement more formally through, for example, school reports and certificates. It is this latter role that is most frequently associated with educational assessment since it is one which is both highly visible and crucial to the whole process of sorting and sifting people for different learning opportunities and occupational roles.

More recently a third function of educational assessment has begun to rival in both significance and visibility these more well-recognised purposes. This is the role that assessment can play as an instrument of *system control.* Assessment is arguably *the* most powerful policy tool in education. Not only can it be used to *identify* strengths and weaknesses of individuals, institutions and indeed whole systems of education; it can also be used as a powerful source of *leverage* to bring about change. Only funding policies can begin to rival the capacity of assessment procedures to define policy priorities and to ensure that they are implemented. It is for this reason that we are increasingly seeing policy-makers in many countries quite deliberately manipulating assessment policies in order to alter the priorities of the education system with a view to improving its effectiveness. Certainly the national assessment programme in England and Wales has this rationale as do many of the so-called "high-stakes" testing programmes in the United States and elsewhere. The more institutional funding decisions or individual career opportunities are governed by the results of a particular assessment, the more

Assessment: Problems, Developments and Statistical Issues.
Edited by H. Goldstein and T. Lewis.

powerful the assessment device in question is likely to be in determining priorities and practices.

Although the deliberate use of MDI (measurement-driven instruction), as it is sometimes called, is a relatively recent phenomenon (Airasian, 1988), the capacity of assessment procedures to influence priorities and practice within the education system is an endemic and enduring feature. Thus, although the scale of the *deliberate* manipulation of assessment policy which is taking place at the present time is a relatively novel phenomenon, arguably just as important have been the *unplanned* consequences which the adoption of particular assessment practices over many years has led to. Arguably, assessment has always been, and will probably continue to be, the single most significant influence on the quality and shape of students' educational experience and hence on their learning. Since the earliest days of mass educational provision, assessment procedures have largely governed the content of the curriculum, the way in which schools are organised, the approach to teaching, and the learning priorities of students. It is both the vehicle and the engine that carries the delivery of education. There is little chance of this situation changing in the foreseeable future. Without some severe hiatus in the link between educational qualifications and employment and unless governments are persuaded that accountability can be provided for in other ways, it is likely that the results of educational assessment will continue to constitute the language by which the achievements of individuals, institutions and even whole educational systems are judged. By the same token, assessment will also continue to constitute the language which informs the choice of ways and means to generate such achievements.

The current ferment in assessment activity also has another dimension. Not only have governments begun to be aware much more explicitly of the powerful potential of assessment policy, they have also become much more aware of the need to make those policies serve a changing international economy; for education systems to move from the preoccupation with finding accurate means of selection which was its enduring concern for the first half of this century and more. Governments are recognising the need to put in place practices which will faithfully reflect the new needs of the workforce for a high level of vocationally oriented skills, competencies and attitudes. While education systems are being pressurised as never before to produce a very high level of achievement for ever larger numbers of students, they are also being increasingly pressurised to retain the best of the old while adding important new dimensions to the curriculum of primary, secondary and tertiary education. In England and Wales, for example, this means science and technology, even for five- to seven-year-olds. In secondary schools it means a strong emphasis on technical and vocational education, including the core skills of literacy, numeracy and information technology, at every level. In higher education, in addition to these emphases, students are increasingly coming under pressure to demonstrate their capacity to apply the skills that they have acquired, in a variety of different situations, and perhaps, most significantly of all, to critique and develop their own practice by an informed ability to engage in self-review. That is, they need to be capable of identifying the action needed to achieve the learning goals they have identified for themselves. It is clear that the worker of tomorrow will need to be capable of undertaking a continual process of self-assessment concerning their current level of skills and competencies and to know in what ways these might be developed through further learning in order better to meet the demands of the workplace. The emphasis on flexibility and adaptability, as well as on a high level of technical skill, which is becoming increasingly characteristic of occupations at all levels (Cassells, 1994), is already being reflected, not only in changing curriculum priorities, but even more significantly perhaps, in the introduction of new elements in the teaching and learning process itself. Children as young as four and five are beginning to be trained in the skills of self-assessment; of portfolio preparation and action planning (Towler and Broadfoot, 1992). As they go through the school system their understanding of themselves as learners and their ability to choose appropriate future learning targets will be enhanced through a variety of opportunities to develop these skills so that they ultimately enter the labour market equipped with the training to operate effectively within it.

But while there are many examples of such developments currently taking place in many countries, the growth of these new assessment approaches is typically being inhibited by the enduring influence of practices rooted in traditional assessment thinking, thinking which is to a considerable extent a hangover from the age of industrialism, when the educational priorities and hence the priorities emphasised by assessment, were very different.

In Britain, a recent report by an independent review body—the National Commission on Education (1993)—argues that, whereas the shape of the labour force was once a pyramid with the great majority of jobs clustering at the bottom level of unskilled manual work, it is now shaped like a diamond in which most jobs are at the middle, technician and white-collar level. Thus, where once the social and economic requirement of the educational system was to find ways which were both accurate and acceptable, of sifting the population in order to identify the few who would fill the occupational positions at the middle and top of the pyramid, the priority now is increasingly to find ways of raising the overall level of achievement within the education system.

It is this very significant change in the occupational structure which lies behind the efforts currently being made by many governments to raise both the overall level of achievement within the education system and to prioritise new competencies and skills. The UK Government's current learning targets which include getting 80% of young people to Level Two of the vocational hierarchy—namely two A levels or the equivalent by the year 2000—is typical in this respect, as is its commitment to incorporating the assessment of core skills at every level. To raise the overall level of student achievement in this way requires two things. First, it is necessary that learners themselves should be both willing to continue their studies and able to raise their level of achievement. Secondly, it follows from this that schools and colleges must be as efficient and effective as possible in order to maximise the learner's chances of success. It is this latter imperative that lies behind the rapid growth in the explicit use of assessment for quality assurance and control, for monitoring and pressurising teachers through a variety of assessment techniques.

Once again, current practice in England may be cited as a typical example in this respect. The national curriculum and assessment system introduced under the 1988 Education Act faithfully reflects the rationale of its blueprint as embodied in the Task Group on Assessment Testing (TGAT) report (Black, 1988), in which National Assessment was conceived as serving four purposes —*diagnostic* and *formative* to support student learning, *summative* to document student achievement, and *evaluative* to provide for accountability. The British government has provided a common framework of curriculum goals against which to measure the achievement of both institutions and individuals. A combination of school inspection; market forces based on league tables of various performance indicators; a range of quality assurance devices requiring the introduction of internal controls; and, not least, financial sanctions, have provided the necessary pressure to ensure that strenuous efforts would be made to achieve these goals within the education system.

But while such combined strategies for raising the efficiency and effectiveness of both students and schools may have a superficial logic, they are fundamentally flawed in that they do not take into account the relationship between assessment and learning. Whether the preoccupation with summative, numerical outcomes concerns individual students, institutional standards or even systems as a whole, the effect in every case is the same—a tendency to assume that accountability, especially when linked with competition, will, in itself, promote better levels of achievement. As Freedman (1995) puts it, "the notion of accountability has become such a priority that few people seem concerned about the knowledge for which one is to account and little attention is given to whether that which is tested has any relationship to the improvement of students' lives". Indeed, as Corbett and Wilson (1990), among many others, rightly argue, in order to improve, any individual or institution must be enabled to identify the source of its weaknesses and use this diagnosis to identify the strategies which are likely to lead to improvement.

Genuine improvement in the effectiveness of learning actually requires a major rethink in the way that assessment is used. This rethink needs to be based on a careful analysis of how and why assessment can promote both individual and institutional learning, coupled with an equally careful analysis of the social and economic priorities that assessment must be designed to serve. Such an analysis is likely to lead to three key insights for future practice. In the first instance, it is likely to lead to a recognition that assessment priorities need to change from an emphasis on accurate *selection and prediction* to an emphasis on promoting and recognising *achievement* in the light of the changing social context of education and hence changing educational priorities. The second conclusion is likely to be that much existing assessment practice is a reflection of these earlier priorities and that new needs require new techniques to be developed. Thirdly, and most importantly, it is likely that such an analysis will highlight the importance of starting this quest from a basis of what is known about how learning—both individual and institutional—is facilitated, rather than doctrinaire assumptions about the value of competition. As Airasian (1988) has argued, the disregard of the effects of centrally derived assessments as reflected in the almost total absence of attempts to evaluate their impact (Ellwein *et al.*, 1988) on students, schools and systems, suggests that the reason for the introduction of such assessments is as much symbolic as real. The motivation for governments to introduce testing programmes, he suggests, may be a symbolic one in that by so doing they are seen to be tackling the issue of standards; that because testing is associated in the public mind with a variety of desirable educational characteristics such as "standards" and "rigour", it is the image rather than any actual effect on learning outcomes that is the major source of their political value. Whether or not they actually improve standards seems, by comparison, relatively unimportant if the amount of research devoted to exploring this question is anything to go by.

In practice there is very little hard evidence concerning the ways in which particular policy approaches to assessment are likely to affect children's learning or even teachers' teaching. So recent is the advent of widespread interest in assessment beyond the powerful, but narrow, orbit of its role in selection, that policy development has substantially outstripped the available evidence from research (Baker and O'Neil, 1991). The last two or three years, however, have seen the generation of quite a considerable range of insights into the more obvious questions in this respect. How do teachers react to the imposition of so-called "high stakes" testing? What is its effect on schools? In what ways may the assessment methods used be criticised? (see, for example, Stake 1991; Gipps 1990).

This chapter offers a further contribution to help inform the debate on assessment policy. It elaborates the core argument concerning how assessment can best be used to promote learning. It reviews some of the more novel assessment approaches which have recently emerged in terms of their rationale and likely impact on learning. The potential of such new approaches to encourage the kinds and levels of learning outcomes which education systems are increasingly under pressure to provide is contrasted with the reactionary and constraining effects of the parallel development of centrally imposed assessment systems whose primary purpose is the uncritical pursuit of accountability and traditional forms of selection.

"Learning is a co-operative venture between learner and teacher", write Harris and Bell (1994), emphasising the fact that learning is strongly influenced by a variety of features of the context in which it takes place. Not only does this mean that, to quote Knight and Smith (1989) "good practitioners" cannot be identified simply by changes in student attainment scores, even though this is a strongly articulated feature of many centrally based assessment schemes. It also carries with it the obligation to construct a theory of learning which takes contextual variables as well as personal variables into account. In his articulation of the socio-cognitive development perspective (SCD), Smith (1991) highlights the importance of addressing the conditions under which learning takes place. An exploration of classroom assessment practices can illuminate both how far such learning is happening and how it is being facilitated. Certain key questions may be identified in this respect in relation to assessment and learning:

1. What is the relationship between quality of teaching, opportunity to learn and assessment in a given social setting?
2. To what extent is opportunity to learn in terms of pupils' idiosyncratic needs reduced by centrally imposed curriculum and assessment arrangements?
3. What is the effect of raising the profile of classroom assessment such that students' achievements are uniformly and comprehensively defined in terms of targets and levels? If teachers' control of students' learning is increased, and teachers themselves become subject to increased control by governments, will this make students less likely to take the risks which are needed to assist learning? Alternatively, will the increased professionalism, awareness and expertise in classroom assessment, which is another spin-off from the high-profile imposition of new forms of centrally imposed assessment (Broadfoot, 1994), serve to enhance learning by making teachers' objectives clearer (Sadler, 1989) and enabling these goals to be communicated more effectively to the pupil?
4. Tharp and Gallimore's (1989) model of learning envisages assessment as having a direct contribution to make to the "scaffolding" of student understanding, both by informing teachers' judgements and the setting of tasks and by enhancing pupils' self-assistance. How far do more centrally generated assessments contribute to, or hinder, this process?

A MODEL OF LEARNING

While there is an enormous research literature on different aspects of learning including, for example, theories of cognition, and meta-cognition, the origins and influence of different kinds of motivation, the effects of the social context and a myriad of different studies of ways of promoting it according to the subject–matter in question, there is relatively little research that links assessment practice with learning (Desforges, 1989). On the one hand, there is a considerable psychological literature which is potentially relevant in this respect. On the other, there is a growing body of evaluation studies which details the impact of a variety of initiatives which are aimed at improving the quality of student learning whether this is in terms of a change in the nature of learning outcomes or an increase in levels of achievement. What is currently missing—and most significantly missing —is an integration of these two areas of endeavour. Apart from some notable exceptions, the very important insights which are available from research into learning are not being used to inform those activities in the area of educational policy and practice to which they directly relate—which have the promotion of more and better learning as their goal.

A long-standing division between sociological and psychological theories has, arguably, led to a tendency to divorce the intra-personal factors affecting learning from those which reside in the social context. Common sense, however, suggests that the two must be closely related and that any adequate theory of learning and of the role that assessment plays in this must be able to embrace both. Among the important external influences on learning will be the following:

1. The specific socio-cultural milieu of the child which might include ethnic background, social class, geographic location and other more specific family circumstances;
2. The school context, including its ethos, location, resourcing, leadership, etc.
3. The individual classroom environment, including the characteristics of the other children, the teacher's style, competence and personal qualities as these determine the overall quality of the classroom climate she creates and the particular characteristics which determine the quality of her interaction with an individual child.

All of these different sources of influence on the social reality experienced by any individual pupil will tend together to exert a major force on the child's approach to, and success in, learning. As Mercer (1991) suggests:

> "The essence of this approach is to treat human learning and cognitive development as a process which is culturally based, not just culturally influenced, a process which is social rather than individual, a communicative process whereby knowledge is shared and understandings are constructed in culturally formed settings. It does not, in principle, oppose the idea of innate elements in cognitive development, but it does suggest that cognitive development is saturated by culture."

The way in which these forces operate is chiefly through social interaction which produces the meanings which are socially patterned and sustained through cultures (Mead, 1934). Among the meanings generated, as Pollard (1992) points out, are understandings of oneself. Each individual thus develops a sense of identity, a sense of a "self" and this awareness influences the ways in which he or she reacts with other people. In short, Pollard (1996) argues, we may distinguish three domains of social factors that influence learning: the intra-individual, the inter-personal, and the socio-historical. The intra-individual domain is the traditional province of psychologists who study the ways in which individuals experience and construct understanding. The interpersonal is the domain of social interaction as studied by interactionist sociologists. It is the area in which meanings are negotiated and through which cultural norms and social conventions are learned. The socio-historical domain addresses the wider context in which learning takes place, its origins and the circumstances of the learner.

Assessment is arguably one of the most important vehicles for creating understanding through interaction. It is through the discourse of assessment in its various forms—oral, written and non-verbal—and its various modes—which include cognitive feedback, behaviour management and encouragement—that individual learners are able to build up a sense of their identity as learners and thus, ultimately, that range of conditioned responses that is manifest in typical types of motivation, conation and hence achievement for any given learner.

Thus, for example, in a comprehensive review of the literature documenting the relationship between assessment and students' learning, Crooks (1988) identifies a range of short, medium and long-term formative effects of assessment in this respect. In the short term, he argues, assessment has a strategic function in the consolidation of existing learning and in highlighting what it is important to learn. It also has a conative effect—where this is defined as the strength of the desire or will to learn—in influencing whether learning is active or passive, and in reinforcing new learning; in helping learners to develop self-monitoring skills and providing feedback and encouragement for student and teacher. In the medium to longer term, it can help to determine appropriate learning goals, influence students' choice of learning strategies, their motivation and their sense of themselves as successful or unsuccessful learners. Crucially, it also influences the student's ability to retain and apply what has been learned.

From Crooks's extensive review, it is possible to distil four broad dimensions which structure the relationship between assessment and learning shown in Figure 3.1. These are: (1) motivation to learn; (2) knowing what to learn; (3) knowing how to learn; and (4) the effectiveness of learning.

MOTIVATION

For every student who is extrinsically motivated by the "carrot" of good results, there are many more that learn not to try for fear of failure. Students whose self-esteem is progressively eroded by negative assessment results frequently acquire "learned helplessness". Their negative view of themselves as learners makes them anxious and less willing to try hard with a task. Research by Bandura (1977) and others shows that the amount of persistence exhibited by students is a function of how successful they expect to be. Previous experience of success or failure is acknowledged to be the prime source of student motivation.

A considerable body of research also highlights the key role of the affective dimension in learning—the way students feel about themselves as learners, their confidence, their enjoyment, their expectations of success, their explanations of failure and the ways in which all these factors can influence their motivation. Mills (1991), for example, distinguishes between motivation arising from a conditioned need for success *per se* and the more natural, intrinsic motivation which is derived from the interest of the task itself. He suggests that the natural motivation associated with pleasure in doing the task for itself—something which is characteristic of young children—often becomes overlaid by learned habits and

Assessment provides motivation to learn
— by giving a sense of success in the subject (or demotivation through failure)
— through a sense of self-efficacy as a learner

Assessment helps students (and teachers) decide what to learn
— by highlighting what it's important to learn from what is taught
— by providing feedback on success so far

Assessment helps students learn how to learn
— by encouraging an active or passive learning style
— by influencing the choice of learning strategies
— by inculcating self-monitoring skills
— by developing the ability to retain and apply in different contexts

Assessment helps students learn to judge the effectiveness of their learning
— by evaluating existing learning
— by consolidating existing learning
— by reinforcing new learning

Figure 3.1 The influences of assessment on students' learning (Crooks, 1988)

self-concepts in relation to learning, such that the existence of motivation becomes dependent on external reinforcements and conditions. Hess and Azuma (1991) provide an interesting illustration of this process in a study which explored the rather different patterns of motivation in children as young as four which are exhibited in Japanese and United States cultures. Whereas the Japanese children were socialised into wanting to replicate the processes they had been told to follow and were satisfied by feedback that told them they were doing this, the American children needed the reward of constant achievement to sustain their motivation for the task.

Such evidence concerning the importance of motivation and the different personal and cultural factors which influence it emphasises the argument made by Raven (1991), concerning the importance of working from pupils' own values if motivation is to be maximised. He suggests (1991, p. 49) that "effective mentors (whether parents, teachers, supervisors or managers) study the values of their children, pupils or subordinates and then create developmental environments in which these children, pupils or trainees can undertake activities which are important to them".

Thus, teachers have to engage in a continuous decision-making process, deciding whether and how a given intervention will help or hinder a particular student's motivation, and hence learning.

DECIDING WHAT TO LEARN

If we look even briefly at what is known about learning, we discover that here also the assumptions underlying most forms of educational measurement are at odds with the insights now rapidly emerging from educational psychology. Constructivist learning theory argues that "Learners become more competent, not simply by learning more facts and skills but by reconfiguring their knowledge. By chunking information to reduce memory loads and by developing strategies and models that help them discern when and how facts and skills are important" (Mislevy *et al.*, 1991). Rote learning, by contrast, is difficult to retain in the longer term.

There is a wealth of research on learning which argues for the use of more diagnostic measurement techniques grounded in a constructivist perspective in which the different levels of student understanding can be identified, as a basis for appropriate instructional intervention (see, for example, Masters and Mislevy, 1991). This perspective contrasts with the more traditional approach to assessment which typically measured outcomes in terms of right and wrong and was not primarily designed to cast light upon students' learning strategies.

A constructivist view of learning conceptualises it as a process in which the individual, striving to achieve an object, encounters reality which they must be helped to interpret through, for example, trial and error or group discussion. In this essentially interactionist perspective, assessment, in the form of formative feedback, is clearly a vital element (see also Ardwoino and Berger, 1986; Perrenoud, 1991).

> "In everyday classroom terms this means teachers using their judgement of children's knowledge or understanding to feed back into the teaching process and to determine for individual children whether to re-explain the task or concept to give further practice on it or move onto the next stage" (Gipps, 1990).

These evaluations are also the basis for teachers' choice of task which they seek to match to the needs of the learner (Bennett *et al.*, 1984).

Yet the process of communication is problematic (see, for example, Abbott, 1996, forthcoming). Harlen and Qualter (1991), for example, argue that feedback tends to be incomplete and fuzzy, and often based on a limited range of potential criteria. Filer (1994) shows how that the assessment situation itself may be constrained in a number of ways such that pupils cannot demonstrate the learning they have got or indeed cannot acquire it. Further, it is clear that the mere provision of feedback is insufficient for optimum learning; it must also indicate what the pupil can do to correct unsatisfactory results (Gipps, 1993).

The difficulty inherent in effecting the communication function of assessment is certainly compounded by the very different perspectives/definitions of the situation, which pupils have, of which teachers may frequently be unaware. Little (1985) provides a useful set of dimensions for analysing these data. Little's study identified 18 categories either from the attribution literature or from the responses themselves, to categorise the explanations which students presented for success or failure. These included "effort", "luck", "past difficulties", "help from others", "mood", "motivation", "time spent", "age" and "facilities" (see also Pollard *et al.*, 1994). She found that "effort", "interest" and "performance ability" were the most frequently used explanations. Three different ability categories, "performance ability", "specific competence" and "general competence" all occurred frequently. As Little concludes, children's explanations of achievement are influenced by a social environment made up of consensual definitions of achievement, achievement values, objective determinants of achievement and teacher and parent beliefs about the importance and determinants of academic achievement. What this analysis does suggest is that children's explanations simply do not reflect an objective classroom reality. Rather they are determined through a complex interaction between objective and subjective reality and that therefore feedback is a very delicate and complex process.

Little's essential argument is that to understand the impact of assessment on children, their learning and their explanations for success and failure, it is essential to start with their own interpretation, bearing in mind too that such explanations are likely to change, particularly in relation to age, but also in relation to other contextual factors. Powerful support for this argument is also provided by the longitudinal research of the British Primary Assessment Curriculum and Experience project's data on pupils' perspectives on assessment (see, for example, Pollard, 1996).

LEARNING HOW TO LEARN

It has already been argued that in today's society, learners need to be helped to develop meta-cognitive skills—to give them the understanding of themselves as learners which will give them the confidence to impose their own meaning on the task and enable them to self-regulate the thinking process. There is plenty of research evidence which testifies to the possibility of teaching even very young children—five- and six-year-olds—the skills of monitoring and managing their own learning (Merrettt and Merrett, 1992; Griffiths and Davies, 1993; Quicke and Winter, 1993). These studies show what can be achieved by deliberate incorporation of opportunities for students to assess their own learning strategies during the process of teaching and learning. Other studies document very similar processes in higher education (Assiter and Shaw, 1993). As Scott (1991) argues, the individuals' ability to self-monitor, to be aware of both their strengths and weaknesses as a learner (meta-cognition) and to be aware of their affective condition is, along with their level of intrinsic motivation and the type of feedback received, a key factor in both task perseverance and relative success in performing that task.

Another crucial aspect of the impact of assessment on learning is the extent to which it encourages or inhibits active learning. Writing of adult learning, Parsloe and Jackson (1992) argue that adults learn best when they are solving problems; when they draw upon their previous learning and experience, when they work collaboratively in groups and their learning is self-directed and therefore "active". Social constructivist learning theory suggests that these four key dimensions would be equally applicable in primary classrooms. At every level, learning, however, tends to be dominated by teachers rather than the learners themselves. It tends, instead, to encourage passivity on the part of the learner. Parsloe and Jackson found that few teachers relish relinquishing power in favour of empowering the learner, especially when they have trained long and hard to achieve this status and have been conditioned to the notion of an active and powerful role. Not surprisingly therefore, as Bachor and Anderson (1993, p. 17) argue, referring to primary schools, "Classroom assessment is an ongoing teacher-centred activity which the pupils seem to be aware of and detached from their own processes of explanation and orientation". Where this is so, it is potentially very significant for as McMeniman (1989) and Decharms (1984), among others, have argued, the degree of control students feel they have over their own learning affects both their motivation and even their actual capacity to learn.

JUDGING THE EFFECTIVENESS OF LEARNING

It is clear, even from this very brief review, that effective classrooms depend on the sensitive and informed use of assessment; that the quality of students' engagement with their learning will be strongly influenced by the way the teacher uses assessment to guide, motivate and reinforce learning. Assessment may be seen as a language, as a form of communication between the teacher and the pupil, the former seeking to identify both achievements and needs, the latter seeking to respond to signals of guidance as to appropriate subsequent activity. But, like any language, it is both delicate and far from perfect as a mechanism of communication. Furthermore, it is a language not only for cognitive dialogue but also one which serves to energise or inhibit learning as it impacts on students' feelings and desire to learn. Hence assessment is a language that is both profoundly complex and vulnerable to misunderstanding, and yet of the very greatest importance in promoting learning. It is a form of communication which requires a very high level of professional skill and understanding to articulate effectively.

Thus, if assessment is to be effective in promoting learning, teachers need to be able to take into account the nature, origins and importance of pupils' orientation to the learning process. They need to recognise that there are different kinds of pupil motivation, the significance of self-esteem in the learning process and the impact of associated factors such as anxiety, the need to achieve and the extent to which students feel that they themselves can influence their level of success in learning. Of particular relevance at the present time is the whole area of pupil explanations of success and failure and how the changing character of classroom assessment as practised by teachers, following government directives, may impact on their evolving picture of themselves as learners and their capacity to function effectively in relation to the learning opportunities provided. The critical importance of assessment as the conduit through which messages are communicated to the individual learner underlines its capacity profoundly to influence pupils' affective state, and hence, to be deeply influential in determining in what way learners will respond to a given learning opportunity. It is therefore important that teachers understand some of the important influences which affect how and why children learn. Certainly teachers' assumptions in this respect, more or less well founded as they may be in terms of psychology, are likely to influence their pedagogic strategies *vis-à-vis* individual pupils. If these assumptions are actually not well matched with the children with whom they are working, then at best there is likely to be a large measure of "interference" in the communication function of classroom assessment which will inhibit its successful use in terms of both directing learners' efforts and in motivating them. At its worst—to put it crudely—it may almost be like the teacher talking German to learners who know only Italian.

And yet while most teachers are well aware of the capacity of assessment to influence children's motivation to learn, their understanding may well not go beyond this. For instance, they may be less able to explain why a given incentive works for some pupils and not others. Equally, it is widely recognised that what is to be assessed will influence what children (and their teachers) put most emphasis on in their learning activity. Yet the precise nature of the relationship between assessment and students' choice of learning strategies is likely to be much less well understood. Even less is generally known about what influences a given individual towards a particular way of engaging with the task of learning and what teaching strategies are most likely to lead to successful and enduring learning.

Whilst, as Bernstein (1977) argued, there are three message systems in the classroom—curriculum, pedagogy and evaluation (assessment), it is arguably assessment that provides the vital bridge between what is offered by the teacher and the reaction to it by the pupil. While assessment may often appear to be a one-way form of communication centred on the transmission of praise or blame, affirmation or correction, by the teacher, a little reflection readily reveals that it can never simply be such. While in the past pupils have been called upon rarely to engage actively in an assessment dialogue, the assumption that they will react to assessment messages in some way is the basic premise on which such activity is founded. Yet there has arguably been a dearth both of research into how and why and with what effect teachers use assessment in the classroom and a neglect of these critically important skills in professional training. Where there has been great attention given to the professional skills associated with classroom management, curriculum delivery and a whole range of other specialist concerns, interest in teachers' understanding of, and ability to use assessment has been largely confined to the keeping of adequate records for the purpose of reporting (Mavrommatis, 1995).

Thus, it is not surprising that in England, for example, Her Majestys Inspectors of Schools (HMI, 1991) wrote: "Existing arrangements for assessing children's performance are inadequate and inconsistent and assessing children's learning is still ineffective in most primary schools." Smith (1991) argues that teachers are still not effectively matching tasks to learners' needs, nor creating the learning environment with sensitivity because they lack a sufficient level of psychological understanding of the kind of variables which impact on the learning process.

De Landsheere (1991, p. 25) offers an even more pragmatic explanation in suggesting that teachers operate on the basis of an intuitive distribution and will adjust the level of their assessment according in order to achieve a distribution—25% weak, 50% average and 25% good—from year to year. Bonniol (1991, p. 129) reinforces this argument in his assertion that

> "the old demon of innate intelligence is creeping in again. It is something to fall back on because no theory of learning is required. When they have had no training in assessment, teachers find it difficult to deconstruct notions of comprehension and problem-solving such that it is rare for them to be able to develop even a sketchy vision or practical model of the cognitive operations they expect of their pupils".

For primary school teachers in England in particular, there has been a deep-seated antipathy to assessment, perhaps because of its association with streaming, selection and the 11-plus (Harlen and Qualter, 1990). The very substantial amount of informal assessment activity which has always taken place underneath the large edifice of formal assessment in the shape of marks, grades, ticks, comments, smiley-faces, tests and examinations, has typically not been seen as "assessment" at all (Harlen and Qualter, 1990) since it has existed to serve quite different purposes and is governed by rather different notions of quality (Broadfoot *et al.*, 1995). Central to the latter, Schwab (1989) suggests, is the criterion of usefulness in relation to curriculum purposes rather than more conventional definitions of validity.

A great deal of energy has been devoted to analysing teaching behaviours in terms of which of them seem most efficacious in promoting learning. Typically such analyses—implicitly or explicitly—distinguish between an instrumental dimension concerning the link between different kinds of teaching behaviour such as the pacing of lessons or the way in which questions are used and the range and quality of student learning and an expressive dimension which, at the most obvious level, embraces the acceptability of particular teacher behaviours for pupils and the manageability of a given approach for teachers. Arguably much more fundamental in this respect than simply the perceived acceptability of given behaviours for teachers and pupils, is the quality of the overall learning climate produced and the impact this has on pupils' learning attitudes to themselves as learners and to the task of learning.

In none of these respects does there seem to have been a sustained or substantial attempt to understand the role of assessment practices. It is as if the traditional location of assessment practices as an essentially summative activity firmly located in the domain of measurement, has prevented teachers seeing their much more informal, often intuitive classroom evaluations as part of the larger whole. Arguably, the traditional conceptual apparatus of assessment involving concepts such as validity and reliability, has operated as a discourse of power effectively limiting the way in which assessment issues were both conceived and addressed. This in turn has tended to limit the way in which the teaching–learning relationship as a whole has been conceived, for, as Freedman (1995) points out, "Models of assessment are similar to any educational model. Such models do not merely describe ideas and practices, they actually work to shape the way people think about and practise education." A similar argument concerning the effects of the domination of a particular assessment paradigm is made by Gipps (1994). But the lack of an appropriate conceptual apparatus to inform thinking about the relationship between assessment and learning is only part of the reason for the relative professional neglect of this topic. Other, more pragmatic reasons are also relevant, notably the structural and organisational arrangements for which assessment information is a vital element. According to Weston (1991) for example, much of the assessment currently practised in schools across Europe is determined by school organisation and systematic pressures such as selection at the end of primary school.

THE WAY FORWARD

Thus a picture emerges of the essential fragility of the learning process and of the key role assessment inevitably plays in it. Learners' willingness to persist in their learning and their ability to learn effectively is seen to be significantly influenced by elements in the interpersonal context which are articulated by means of the assessment language.

It has also been suggested that teachers' professional understanding in relation to assessment has typically been one of the least well-developed aspects of their expertise. Apart from some significant exceptions, teachers have tended to adopt a pragmatic approach to assessment at every level of the education system which involves a relatively uncritical acceptance of tried and tested procedures for both formative and summative purposes (Scarth, 1984; Heywood, 1994; Mavrommatis, 1995).

Now, however, as discussed earlier in this chapter, teachers are being faced with new challenges, both to raise the level of their students' achievements and to extend their range of competencies. Furthermore, they are being faced with new mechanisms for accountability which increasingly locate them as the *receivers*, as well as the initiators of assessment judgements. It is therefore important to question how far and in what ways these pressures are effecting a change in teachers' assessment practices. Thus the final part of this chapter is devoted to a discussion of the ways in which assessment needs to develop to promote more relevant and effective learning which will meet the needs of the twenty-first century and the extent to which such developments are, or are not, taking place at the present time.

In recent years, the distinction between summative assessment purposes which have as their rationale the identification—and often the communication of learning outcomes—and formative assessment purposes which have as their rationale the use of assessment to support learning, has been widely recognised (see, for example, Torrance, 1993). But while there is growing recognition that both these purposes are important and that they may well require different techniques, there is little sign as yet that this recognition is eroding policy-makers' preoccupation with measurement as the key to higher standards.

If an explicit concern to link assessment practices with research-based insights concerning learning is relatively recent (Broadfoot, 1994; Gipps, 1994; Torrance and Pryor, 1995), more pragmatic attempts to respond to the new challenges facing the education system have burgeoned in recent years. The United Kingdom has been typical of many other countries in its attempts to broaden the basis for certification by the introduction of new and more flexible modes of examining. These attempts have included course-work assessment, modular and unit accreditation, and locally based syllabus development, all aimed at overcoming the perceived shortcomings of traditional examinations by providing better opportunities for students to demonstrate their achievements (Murphy and Broadfoot, 1995).

The work of Desmond Nuttall, which is the subject of the Murphy and Broadfoot volume, illustrates more clearly than perhaps that of any other scholar the changing twists and turns in the search for improved assessment practices. It documents, in particular, the distinction between, on the one hand, reforms which have been focused within the prevailing commitment to examinations, and on the other, the challenge to examinations *per se* as an effective and useful assessment device. Nuttall's work also aptly traces the significant impact of the growing preoccupation among policy-makers with measurement and the associated paraphernalia of performance indicators and league tables.

The history of assessment development during the last three decades is a rich and confused one in which only the steadily growing importance of assessment is not in dispute. The wealth of initiative does, however, document four important themes. One of these is the search for greater accuracy in measurement in the interests of both efficiency and social justice. A second concerns the quest for raising levels of student performance by providing assessment mechanisms with more power to motivate students through more flexible curriculum provision and an emphasis on success rather than failure. A third theme centres on actually empowering learning by means of assessment and the fourth and final theme concerns changes in who has responsibility for designing, conducting and using the assessment. These four themes are neither clearly separate nor comprehensive. However, they may serve briefly to illustrate the overall argument of this chapter concerning the need for assessment thinking to address what is known about learning if education systems are effectively to address the economic challenges of the late twentieth century.

In terms of the first theme, a great deal of energy has been devoted by the assessment industry in recent years to addressing issues such as removing cultural and gender bias from tests and examinations, the provision of multilevel papers and refining techniques of ensuring comparability and reliability. In the United Kingdom, the United States and elsewhere, there has also been a significant growth of concern with validity, heralding the advent of a search for more "authentic", curriculum-integrated, tasks into the range of assessment techniques like some of the English Standardised Assessment Tasks (SATs) and the United States "New Standards" project (Resnick, 1991). But these developments, while introducing important and fundamental new perspectives on the purposes and priorities which should inform the design of assessment procedures, have also raised many new technical problems. It is clear that the more assessment is designed to address the full range of learning goals, the more it is designed to be humanistic rather than technicist, the more difficult it is to ensure an acceptable level of reliability where summative judgements are concerned. Such developments—and the technical challenges associated with them—are clearly closely linked with those of theme two which relates to initiatives designed to raise student achievement by recognising what is known about the relationship between assessment and learning. Such links are well evidenced in the currently widespread international initiatives to introduce continuous, course-work-based assessment to complement, or even to replace, more conventional examinations. (Broadfoot *et al.*, 1990; Mauritius Examinations Syndicate, 1994).

Criterion-referencing has clearly made a great impact in relation particularly to theme two in its emphasis on the mastery of explicit goals and a corresponding reduction in the encouragement of competition between students so endemic in the traditional "measurement" model of assessment. This change of emphasis has made possible a wealth of new assessment approaches emphasising competence of which the new vocational qualification frameworks in the UK, Australia and New Zealand provide important examples (Wolf, 1994; Curtain and Hayton, 1995). The increasingly pervasive language of modules and units, accreditation and curriculum frameworks heralds a new integration in the way curriculum and assessment are being thought about internationally, in which the emphasis on making learning goals explicit and encouraging students to plan their own learning paths in terms of them is likely to be of considerable importance in raising overall levels of student achievement.

This is not to say, however, that moves towards "standards-based assessment" are not fraught with many problems both developmental and generic. The many recent critiques of criterion-referenced approaches to assessment highlight both the fundamental technical difficulties of defining and implementing standards unambiguously and the practical difficulties of designing qualification systems based on this approach which are both rigorous and flexible (Wolf, 1994).

The graded tests movement which enjoyed brief popularity in England in the 1970s and 1980s provides an excellent illustration of the rationale behind much of the more recent thinking in assessment reform. They were very influential in helping to establish the importance of at least two of the factors which we have seen linking assessment and learning—motivation and confidence—firmly on the professional agenda. However, they were also difficult to manage in practical terms and subject to significant conceptual and technical difficulties—not least the identification of unidimensional domains of achievement and the identification of appropriate levels of challenge to provide comparability between the various grade levels (Goldstein and Noss, 1990). These difficulties have been well illustrated in the subsequent incorporation of the graded-tests rationale in the 10-level, criterion-referenced design of the English National Curriculum Assessment system in which it has proved to be very difficult both to identify unambiguous gradations of difficulty in each subject and to design suitable assessment tasks to elicit performances on which these can be judged. Nevertheless, it remains the case that learning, and how it can best be fostered, is a significant driving force in such initiatives, an emphasis that represents a significant departure from the traditional domination of concerns with the accuracy of measurement in assessment discourse.

Such a transformation in assessment policy priorities has yet to be achieved in relation to theme three. Certainly there is a rich variety of self-assessment initiatives currently being developed in many primary, secondary and tertiary education contexts around the world. These include the use of student-managed portfolios, learner diaries, records of achievement and other similar initiatives being developed by teachers with the explicit intention of training the students' meta-cognitive strategies and awareness so that the student becomes empowered as an autonomous learner. Nor does the impact of this thinking stop with the individual learner. Insights about institutional learning which have become associated with notions of the "learning organisation" place equally heavy emphasis on the key role of assessment procedures in empowering institutions to become willing and able to learn and improve themselves. If, in the mid-1970s, "self-assessment" was a term without an intellectual pedigree (Broadfoot, 1979), it has become one of the core "life-skills" for the 1990s.

Such a massive increase in acceptance and use is testimony to the power of self-assessment to promote learning wherever its use is based on an understanding of *why* it works. As the earlier discussion of social constructivist learning theory made clear, self-assessment provides the necessary opportunity for reflection on the part of the learner and for discussion with the teacher that facilitates the provision of appropriate "scaffolding"; it emphasises the unique perspective of the individual learner; and the importance of both affect and cognition in effective learning. It is an externalisation of the process of meta-cognition and a way of training these skills.

The contrast between self-assessment in all the many guises in which it is finding expression and the current policy emphasis on centrally imposed standards highlights the ultimate importance of theme four for assessment and learning. A great deal of the responsibility for designing, conducting and using assessment has passed out of professional control in recent years and into the hands of governments anxious to use this powerful tool to exact change within the system. The increasing politicisation of assessment activity has inevitably tended to prioritise the measurement agenda over a potentially developmental one. Just at the time when major advances in assessment thinking were leading to what might have become profoundly important developments in the management of teaching and learning, the measurement paradigm has been reincarnated with renewed vigour in order to satisfy politicians' apparently insatiable appetite for reassurance over "standards".

This is not true of all countries—some, like Scotland and France, seem to be resisting what they perceive to be the damaging effects of such discourse and its emphasis on results rather than causes; on refining measurement rather than promoting learning. Either way, government interest in assessment standards in education systems around the world is a potent force to help or hinder change. As the introduction to this chapter made clear, assessment has long been one of the most powerful influences shaping educational provision. Hitherto, much of that shaping has been incidental. Now it is increasingly deliberate at a time when the implications of different assessment paradigms are becoming ever more clear.

It has been the argument of this chapter that a teaching force which is professionally informed and trained concerning assessment has considerable potential to promote more effective learning. Where teachers and institutions find themselves trapped in the rat-race of league-tables and other centrally imposed assessment measures, the window of opportunity is likely to close as swiftly as it opened. This is well illustrated in California, for example, where the unprecedented involvement of teachers in the marking of state-wide tests led in turn to an upsurge of interest in the whole subject of assessment and how it could be used to promote learning. A development that is in many ways similar has taken place in England among primary school teachers. Hostile to the centrally imposed national tests, this challenged them to find an alternative, more professionally acceptable response to the demand for more explicit assessment on the part of parents and the government. The result is that many of them have emerged with a new understanding of, and enthusiasm for, the potential of assessment as a learning tool (Broadfoot *et al.*, 1995).

Education at the present time is the site of many struggles over priorities. Disputes about funding and provision; quality and professionalism; changing pupil needs and social priorities are currently figuring prominently in the education policy debates of many different countries. But perhaps the most important, and yet the least recognised, is that between the different purposes of assessment. If governments are genuinely interested in raising standards, they would do well to recognise not only the power of assessment as a policy tool which can be used to impose priorities and provide for accountability. They need also to build on the lessons of the research and development that has been taking place in recent years in the field of educational assessment and the now extensive experience that exists among teachers and other educational professionals concerning the way in which the power of assessment can be harnessed to help realise the educational priorities of the next century. This means engaging with new technologies of assessment; with the new philosophies and priorities that are currently emerging just as policy-makers did a century or more ago when major changes in the economic and social structure of the industrialising countries also called for a fundamental reappraisal of existing practice. Above all it means recognising that progress in this respect cannot be achieved by the arbitrary imposition of central control. It can only be achieved by the active, professionally informed collaboration of teachers in the generation and implementation of assessment policy priorities.

CHAPTER 4

Statistical and Psychometric Models for Assessment

Harvey Goldstein

Institute of Education, London, England

MODELS FOR COMBINING INDICATORS OF UNDERLYING ATTRIBUTES

Francis Galton (1884) was perhaps the first writer to introduce the notion of mental "traits". By making an analogy with physical size he supposed that there were dimensions of the mind and that the task was to find good ways of measuring such constructs. Much of twentieth-century psychometrics has devoted itself to this task and in the process has spawned a large literature. Much of this literature is highly technical, but its effect has been immense and it has influenced the conduct of almost all forms of quantitative assessment. Its historical development can be traced from the early attempts at formulating factor analytic models (Spearman, 1904) through "item analysis" procedures for designing mental tests (Gulliksen, 1950) to "item response models" (Lord and Novick, 1968) and Bayesian inference procedures (Mislevy, 1994). Most strikingly, all these developments have adopted as an underlying assumption the existence of one or more psychological attributes, and have supposed that a set of observed responses can be used to provide evidence about the state or value of these attributes. To keep matters simple we shall for the most part assume that interest is in a single-dimensional attribute, namely where a single "parameter" is a sufficient description of any individual subject's underlying attributes. It also happens that this is the assumption, however inadequate, behind most activity in this area.

The raw material for the psychometric model consists essentially of a set of indicators, typically yes/no responses to test questions or items. They may also be parts of complex questions or components of free form responses, essays, or components collected from across a much larger set of responses. They may consist of an ordered grade or mark scale, but this raises no fundamentally new substantive issues. These indicators are all assumed to reflect, in various degrees,

Assessment: Problems, Developments and Statistical Issues.
Edited by H. Goldstein and T. Lewis.

the underlying attribute which we wish to measure, whether this be reading achievement, intelligence or logical reasoning ability. For convenience we shall use the term "test" or "test score" generically to refer to any measurement whether it be along a continuous scale or simply a binary success/failure grading.

The first issue to confront a would-be test developer is therefore that of deciding which indicators to select. Clearly this is a non-trivial task and one where there may indeed be no agreement. Moreover, the set of indicators generally will be contingent upon the particular context in which they are collected, or the test administered. Thus, a set of items designed to assess understanding of a particular mathematics course generally would not be expected to be appropriate for the assessment of a quite different course. Apparently innocuous features of assessments, such as the layout and language used in questions, can strongly influence responses (see for example APU, 1985). We shall have more to say about such matters below, but for now suppose that we accept a set of item responses as indicators and wish to use these to place a value on the underlying attribute.

What is required is some form of summarisation. If the items are all yes/no responses then the simplest procedure is to count the proportion of correct (yes) responses and use this as the required measure. It may be, however, that we wish to weight some items differentially, perhaps because we think they are more relevant to the underlying attribute than others, or we may wish to give less weight in some cases to "difficult" items that "weak" subjects get correct if we think that "guessing" has taken place. It is important to note that there is no necessary requirement here to suppose that individuals possess only a single attribute. Thus, we may be interested in achievement in algebra, but recognise that our items to some extent reflect arithmetic, language competence, etc. The point is to choose a set of items all of which reflect the attribute chosen and then to combine them suitably. We would also need to pay attention to the relative weights of different types of algebra item. Thus, for example if we had available a very large number of simple one-variable equations to solve then we might wish to allocate proportionally less weight to these as opposed to a fewer number of items dealing with pairs of two variable simultaneous equations. In such a case we are recognising that even within algebra it is possible to distinguish further topics or "dimensions" of interest, so that what we define as a dimension of interest may be regarded as a convenient heuristic. We also need to recognise that, having done this, we have set up in effect a *definition* of the attribute concerned. Procedures such as this are common, for example in constructing price and wage indexes, and their properties and limitations are fairly well understood and accepted.

Psychometric theory has proposed an additional, and crucial step. Namely that with some additional mathematical assumptions, principally about dimensionality and conditional independence, it becomes unecessary to define the dimensions explicitly as above. We may use data from sets of observed responses to derive estimates of the underlying attributes from an *internal* analysis of the data. We look first at the assumption of dimensionality.

THE DIMENSIONALITY OF A SET OF ITEMS

Consider a simple model which expresses the *probability of a correct response* to each item as a simple function of one underlying attribute

$$\pi_{ij} = \alpha_i - \beta_j \tag{4.1}$$

where α_i, β_j respectively can be interpreted as the "ability" of the ith individual and the "difficulty" of the jth item or question. The actual observed response takes the values 0 (if incorrect) or 1 (if correct), and in the simplest case we can assume that these responses have a binomial distribution, conditional on the set of probabilities, or equivalently the parameters in equation (4.1). We shall return to this latter assumption shortly. We suppose that the underlying attribute is continuous.

From a statistical point of view we have some further choices available concerning the parameters of this model. In particular we can impose distributional assumptions upon them, for example that the α_i are drawn at random (or exchangeably) from a population, in which they are distributed, say, according to a normal distribution. If our individuals are selected at random then this may be a reasonable assumption. In such a case equation (4.1) is then a special kind of factor analysis model with a single factor and a binary response. We might also impose a distribution upon the β_j if we considered that the items were sampled from some "superpopulation" of items. This would lead to so called "generalisability" models (Cronbach and Webb, 1975). Further elaborations of equation (4.1) would be to add covariates, for example we might suppose that response probabilities were a function of social or educational background, gender, ethnic origin, etc. We might also suppose that the between-individual distribution depended on such factors so that we would allow the (normal) variance in the model to depend on the covariate values. For such purposes it is convenient to view these one-dimensional models as "multilevel", for example in this case as two-level models where items (level 1) are "nested" within individuals (level 2) (Goldstein, 1995).

We can extend models such as equation (4.1) in a further direction by allowing the relationship between the response attribute and the ability to vary across items. Thus we can write

$$\pi_{ij} = \gamma_j \alpha_i - \beta_j \tag{4.2}$$

where the γ_j, one for each item, are usually known as discrimination parameters since they refer to the rate of change of response probability with ability. They function as weights in the estimation of individual abilities and are analogous to the differential weights we might wish to apply to our set of indicators when aggregating into an overall score. In the context of a one dimensional model they are estimated from the data. Because the model assumes a single dimension an item that has a low association with most other items will tend to have a low value, etc.

A difficulty with equations (4.1) and (4.2) is that if the attribute π_{ij} is regarded as a probability, it allows the prediction of probabilities outside the range (0,1). To deal with this we can adopt the standard procedure of inserting a non-linear "link" function and write

$$
\begin{aligned}
g(\pi_{ij}) &= \gamma_j\alpha_i - \beta_j \\
g(\pi) &= \log[\pi/(1-\pi)] \quad \text{logit} \\
g(\pi) &= \log[-\log(-\pi)] \quad \text{log-log} \\
g(\pi) &= \pi \quad \text{identity}
\end{aligned}
\tag{4.3}
$$

where the identity link leads back to equation (4.2). The interpretations of all three models and their elaborations are essentially the same. It is interesting historically that the developers of the logit link function model made little connection with previous procedures based upon the identity link model.

We note that the parameter estimates we obtain from applying the different models, in particular their rankings, will not in general be the same (Goldstein, 1980). In this sense the choice of link function, that is, the choice of the exact form of the relationship between response probability and individual "ability", known as the "item characteristic curve", is part of the definition of the measurement scale and will directly affect our inferences. This property is shared by all other "latent variable" models where there is a choice of link function.

The parameters of any given model will in general depend upon the population from which the individuals are drawn. Because such factors as social background and curriculum exposure may be expected to affect the item responses in complex ways, the parameter values and their rank orderings may also be expected to vary across populations. Thus, estimates obtained for one social or ethnic group may not apply for any other group. Likewise, estimates obtained from a representative sample of the whole population of a given age or grade being educated within any one educational system will necessarily be averages over all possible subpopulations and hence not necessarily appropriate for particular subpopulations. We shall return to this issue of population dependency below (see also Chapter 6).

If we choose the identity link and assume that each α_i is a separate parameter to be estimated, then a natural (although not fully efficient) estimator of an individual ability is just the total score, that is, the number of correct responses (Goldstein and Wood, 1989). For the logit link an efficient (maximum likelihood) estimator in this case is just a monotonic (order preserving) non-linear function of the total score. For the log-log–link, however, the maximum likelihood estimator is not such a simple function and in fact the ranking of individuals is not necessarily preserved (Goldstein, 1980)! The choice of identity link leads to a set of procedures known in general as "item analysis" and the choice of logit link leads to the set of procedures, most commonly in use today, known as "item response theory". There appears to be little attempt by psychometricians to explore other link functions.

It is worth noting that the arbitrariness involved here does not generally apply to observed variable models, such as generalised linear models (McCullagh and Nelder, 1989) since there the measurements are defined before they are incorporated into the model, whereas in the present case the statistical assumptions effectively define the measurement and its scale properties.

We see, therefore, that the formalisation of our original problem of summarising a set of indicators leads to a wide range of possibilities for statistical modelling. It also raises a number of difficult issues about the assumptions which are appropriate and we shall now look at the second crucial assumption, namely that of conditional independence. Following that we shall return again to dimensionality.

CONDITIONAL INDEPENDENCE

Conditional, or local, independence is simple to define. For any link function, we write the observed response (with denominator 1)

$$y_{ij} \sim Bin(1, \pi_{ij})$$

and in addition we assume that, given the π_{ij} these responses are independent. That is, the responses are conditionally independent given the model parameters. This conditional independence assumption is basic to almost all latent variable models. In principle it is possible to test it, but only if we could carry out multiple independent applications of giving the same set of items to the same set of individuals. Since this is usually impossible in practice, some form of conditional independence assumption seems necessary. The assumption of conditional independence underlies a large number of common procedures, for estimating reliability as well as dimensionality explorations. Substantively it means that for a given individual, that is for an individual with particular model parameter values, the actual response to an earlier item on a test has no effect on subsequent probabilities of correct response. It implies, therefore, an essentially static model of individual response. An alternative model, for example where an individual may undergo learning during the course of responding to a test, would not fit this pattern.

MULTIDIMENSIONAL MODELS FOR ITEM RESPONSES

Suppose we have a collection of items as indicators of mathematical ability and we wish to aggregate them into two indexes, say one for algebra and one for arithmetic. If we were prepared to assign separate weights to each item for algebra and arithmetic then we could form two separate weighted averages to give us the required indexes.

In model terms we require a two-dimensional model, a simple example being as follows:

$$\pi_{ij} = \gamma_{1j}\alpha_{1i} + \gamma_{2j}\alpha_{2i} - \beta_j \tag{4.4}$$

where the probability of a correct response is now determined by two individual level parameters. These would typically be assumed to have a bivariate normal distribution and are often assumed to be orthogonal. In contrast with the conditional independence assumption which was entirely concerned with the intra-individual, between-item-response, variation, the dimensionality assumption is concerned entirely with the inter-individual variation.

If we have a collection of responses to a set of items from a suitable sample of individuals, then we can explore the dimensionality. For details see, for example, Bock *et al.* (1988). To do this, however, we have to assume, for any given model, conditional independence. In this sense the two assumptions, about conditional independence and dimensionality, are linked and in practice it would be extremely difficult to test either in isolation. In addition, of course, we must choose our link function and it is quite possible that what may appear to require, say, three dimensions with a logit link function will only require two with a log–log function if the latter is a better representation of the data structure. In addition, as has already been pointed out, we may require different parameters and hence different dimensionalities in different subpopulations.

To illustrate the importance of the dimensionality assumption suppose that we have a set of item responses where there are two dimensions and the inter-individual variation is bivariate. Suppose now that we make the assumption that in fact there is only one dimension and fit such a model, assuming conditional independence. We will find, given a suitably large sample, that the model does not fit the data. One course of action is to elaborate the model, for example by fitting a further dimension or, for example, relaxing the conditional independence assumption (see Jannarone, 1986). An alternative approach which has been advocated, for example by Lord (1980), is that items should only be aggregated or scaled together when they can be demonstrated to be one-dimensional. This view in some ways resembles that adopted by proponents of "criterion referenced testing" which we deal with below. In particular it regards unidimensionality as an essential condition for adequate measurement. In our case this would lead the analyst to select only items from the set which satisfied this condition. In general, of course, this strictly will not be possible since all the items may be expected to have non-zero values for each of γ_1, γ_2 although in practice some of these may be small and lead to some items being discardable. If there is a substantive rationale for discarding particular items on the basis of content, etc. then the procedure of fitting and examining a unidimensional model may be helpful. The fitting of such a model and discarding of items in order *simply to produce a unidimensional structure* provides little further information since it will be highly sensitive to the decisions made by the test constructor when the original set of items was devised.

For example, if a collection of items largely reflecting algebra is mixed with a very few items largely reflecting arithmetic, then the fitting of a unidimensional model will tend to eliminate the latter since these will be weakly associated with the dominant dimension of algebra. The unidimensional outcome is therefore determined by the decisions of the test constructor about the balance of items to be considered and the model fitting process does little more than reflect this. In other words, the role of these models is essentially *confirmatory* (in the language of factor analysis).

THE ROLE OF TEST ITEM MODELS

We see that the model-fitting process relies heavily on several assumptions, notably those about dimensionality and conditional independence. Where these can be satisfied then these models provide powerful tools for design and analysis of quantitative assessment instruments. The extent to which this actually occurs is a matter for debate, but in any case the results of the model fitting process are still heavily dependent upon substantive views about the relevance, balance and detailed format of items and questions. Particular models, such as the logit or identity versions of equation (4.1), lead to test scales which are equivalent to certain simple aggregated and non model-based indexes. This does not imply that we can therefore justify the use of a particular model because it produces results which coincide with a particular aggregation method, nor does it justify that aggregation method. Nevertheless, model-based estimates of individual scores are all particular linear or non-linear averages over the item responses. If we are prepared to accept the model assumptions then, for the items given and the population being sampled, the individual scores can be estimated from the data. Because the assumptions we need to make are strong ones, once accepted the models become powerful instruments for solving a number of problems.

We now look at some particular applications, notably in test equating (see Chapters 5 and 10) and reliability estimation. We shall then look at models where the underlying distribution of ability is assumed to be discrete rather than continuous.

TEST EQUATING

A common requirement in education and other fields is for different measuring instruments which are ostensibly attempting to measure the same attributes or collection of attributes to be calibrated against a common "standard". Thus, the Scholastic Aptitude Test (SAT) used in the USA as part of university admissions procedures needs to be updated regularly with new items, yet at the same time there is a requirement for "equivalence" of mark scales over time. By *equivalent* is meant that any of the alternative forms of the SAT could be substituted for each other without changing the rank ordering of the individuals who take it. Of course, it may not be possible to achieve this, even in principle, since any change in the test may affect the ranking of individuals who take it. Thus, a measure of "equatability", an *equatability index*, can be used to indicate the degree to which rankings coincide, for example measured by the percentage of all possible pairwise comparisons among individuals where the rankings are reversed for the two tests. Technical details of various equating procedures can be found in Holland and Rubin (1982).

The two most common procedures for equating designs are as follows:

1. Where separate samples are drawn from the same population and each one is given one of the tests to be equated.
2. Where all the tests, or more commonly two at a time, are given to each individual in a sample from the population of interest.

In both cases a functional relationship between the test scores is established and this is used subsequently to convert scores from one test scale to another, or to some common standardised scale. The second design is more useful since, with suitable balancing of administration procedures, it can supply an estimate of the equatability index and so tell us how reliable is the equating procedure.

The first problem with equating procedures is that they are population specific. Two tests or assessments may have a high index of equatability in one population but a low index in another. Likewise, within a large heterogeneous population there may be well-defined subpopulations with quite different equating functions. This latter situation will pose a serious problem of choice: namely which population/subpopulation to choose as the reference one. For example, if for two versions of the SAT it was found, say, that equating functions differed by ethnic group, it would be possible to argue that each group should have its own equating procedure. To do otherwise would raise difficult issues to do with how to produce some kind of "average" equating function. If this were done simply by using the whole population this would be equivalent to averaging with weights dependent on the proportions of each ethnic group in the population, with the possibility that this may disadvantage minority groups. Yet to have separate equating functions would inevitably lead to inconsistencies since there would almost certainly be examples where a member of one ethnic group scored higher than a member of another group yet received a lower equated score!

If we extend the use of the term "population" so that it is defined with respect to time and age we encounter the same range of problems when attempting to equate from one time to another or from one age to another. The latter is sometimes known as "vertical" equating and the idea is that we require a common scale for a wide range of ages of children where we cannot use a common test or assessment. Thus, for example, if we had a test of reading for 7–10-year-olds and a different one for 9–14-year-olds we could use a sample of 10-year-olds to equate the tests. The problem here is that while the equating function may work reasonably well, its extrapolation to higher or lower ages would have little empirical support. An alternative procedure would be to use an intermediate age test, say appropriate for the age range 8–12-year-olds to provide a "chain" of equating functions.

Likewise, if we wished to equate two tests a year apart in time, a common procedure is to give both tests to a sample six months after the first was used and to use the resulting equating function. This may work reasonably well when the time gap is not too large, but if repeated over a longer period of time it will be difficult to judge whether the final equatings across the longer time period are acceptable. This illustrates a more general point about the problems of comparisons over time, namely that if the population changes we cannot guarantee that any equating relationship will continue to apply. Populations can certainly change in many ways over time; and in the context of educational assessment, since the curriculum generally changes over time, so the understandings, skills and knowledge of students will change and any attempt to use an equating relationship derived from an earlier population will distort the results (Goldstein, 1983).

ITEM BASED EQUATING PROCEDURES

The above discussion has been solely in terms of the equating of overall test scores, however derived. There are, however, procedures which are based upon item response type models of the kind we have discussed.

Suppose that we have available a very large set of items from which we can choose any subset to form a test. Such a collection is often referred to as an "item bank". If we sample items at random each time we wish to construct a test then, on average, each test will have the same characteristics, and with some assumptions about the relationships between test length and score (which could be verified empirically), it is clear that we can set up a satisfactory equating procedure. A variant on this is to apply item response modelling procedures to such a bank so that, for each item, we estimate its parameters, say, in a unidimensional logit model. With a knowledge of these parameter values we can then select any set of items, not necessarily at random but perhaps tailored to a particular level of difficulty, etc. and still be able to calibrate and hence equate the resulting tests. Indeed, item banks are sometimes promoted on the grounds that anyone can select any set of items to suit their own purposes and still have a common scale against which to equate their results. Thus a school could select just those items designed to reflect their own particular curriculum approach and still claim to be able to make general comparisons. We note, however, that such a procedure requires all the assumptions we have already discussed for item response models, so that its use must remain somewhat problematic.

Another closely related development is that of computerised adaptive testing. Here individuals respond to an initial small set of items drawn from a large calibrated pool or bank. From these initial responses an estimate of ability is made and further items selected which are appropriate for making an efficient estimate for an individual with such an ability, for example in terms of their difficulty levels. This process proceeds until a sufficiently accurate estimate is obtained. Clearly, such a procedure relies heavily upon the assumption of conditional independence and upon the particular chosen dimensionality of the item bank—typically unidimensional.

A major difficulty with the item bank idea concerns whether it is possible to maintain a collection of items whose properties remain invariant over time. We have already remarked that, in education at least, the concept of an item with constant properties is problematic in the light of changing contexts. It is also worth remarking here that "context" applies not only to the external context but also to the internal test structure. Thus, the US National Assessment of Educational Performance found that the properties of items appeared to be affected by where they actually appeared in a test (Beaton and Zwick, 1990).

There is a weak form of equating, sometimes referred to by the term "comparability" and which is still sometimes used within the English and Welsh public examination system (FitzGibbon and Vincent, 1995). This attempts to equate scores for ostensibly dissimilar attributes, for example mathematics and French, using a third "anchor" or "reference" test. This should perhaps not be regarded as a form of equating at all since it is difficult to see how, starting from an assumption that the separate attributes do not coincide, there can be any rationale for attempting to equate them.

Finally, we have been concerned exclusively with statistical procedures for equating. In Chapter 5 Cresswell tackles the issue of judgmental based procedures and explores the social basis for these.

CRITERION REFERENCING

The procedures we have discussed so far have been concerned with making inferences about continuous underlying attributes from the evidence presented by a set of responses to questions and items. The attributes themselves are considered to vary among individuals and that variation can be modelled. Criterion-referenced testing or assessment is concerned with the identification of narrowly defined "domains" or "criteria" and the use of observed responses to determine whether any particular individual satisfies a criterion or belongs to a domain or, for example, has "mastered" a topic. Wolf (1993) provides a discussion and general critique of criterion referencing. As with continuous attributes there are two general approaches to the estimation.

The first approach, as before, is to use an aggregation procedure over a set of observed responses. For example, if we wish to decide whether a student has mastered the topic of "solving quadratic equations" we would observe responses to a set of questions or items, or perhaps a more informal collection of classroom observations by a teacher. One aggregation procedure is to require every question about quadratic equations to be answered correctly. More generally we can allow judgements about the probability of having mastered a topic so that the number of correct answers will be taken into account. Of course, once we admit this possibility we are effectively back with a continuous underlying attribute! Thus a major problem with criterion referencing is that it appears to require a single yes/no decision if it is to remain distinct from all the procedures we have described in preceding sections. There is an intermediate position whereby there could be a small number of classes into which an individual can be classified. This does not really resolve the problem, however, since there will still be probabilities associated with the classification into any one class.

The reason why we need to consider probabilities of classification (and misclassification) lies in the inherent uncertainty attached to any single indicator. By definition, the value of an indicator is not isomorphic to the underlying attribute it is meant to reflect. In other words, it measures it with uncertainty. This uncertainty arises from a number of factors. We have already referred to "context" defined broadly as the external environment and the internal structure of a test or assessment. There is also the issue of the actual choice of real life context to which an item might refer and the format of the item itself, all of which can influence the response (Murphy, 1989). In addition, we are usually faced with a very large, perhaps infinite, number of possible items from which to choose. Thus uncertainty is inherent, and although it will generally be reduced if a large number of items is used it can never completely be eliminated. Hence, for example, if we are to use the criterion of getting every example involving quadratic equations correct, we may well find that with a different choice of examples, perhaps involving different sets of numbers, formats or wordings and perhaps administered at a different time of day in a different physical environment by a different person, then an individual may well sometimes get them all correct and sometimes not.

Wolf (1993) points out that attempts to deal with this problem inevitably lead to the specification of more and more narrowly defined domains with the end result that they are no longer useful as assessments. She also points out that even if it were possible to provide unequivocal classifications for a number of criteria, there would remain serious problems about interpreting a potentially large collection of narrowly defined criteria. Very many users would wish to have some summary measure, but to provide this would undermine the purpose of providing the separate criteria. Furthermore, for the majority of individuals who satisfy only some of the criteria, any summary measure cannot be unique in the sense that a given summary score or grading could be achieved in general through different patterns of responses.

Model-based approaches to criterion referencing parallel those for continuous underlying attributes. In the simplest case we require a model which relates the probability of a correct response to an indicator of category membership, say whether or not the individual is a "master". We write for the ith indicator for the jth individual the probability of a correct response as follows:

$$\pi_{ij} = \pi_i \pi_{mj} + (1 - \pi_i)\pi_{nj}$$

where π_{mj} is the probability of responding correctly if a master; π_{nj} is the probability of responding correctly if a non-master; π_i is the probability that the ith individual is a master.

There is a similar probability for an incorrect response. If we make the assumption of conditional independence as before then the joint probability of the entire set of responses is simply the product of such terms. From this we can then obtain, for example, maximum likelihood estimates of the parameters in the model. As in the continuous case we can assume a population distribution for the individual probabilities, for example that these are derived from an underlying normal distribution.

A generalisation is to suppose that there are several underlying true categories, not necessarily ordered, leading for example to so-called "latent class" models. The model can also be extended so that individuals are classified into a cell of a cross-classification of categorised attributes, so allowing more than one dimension. Note that individuals are classified in these models in terms of their probabilities of belonging to a category so that we can interpret the underlying attribute as a continuous one if we wish.

As with continuous attributes we see that these models share analogous assumptions and are perhaps best viewed as just a special case of the former kinds of models.

NORM REFERENCING, RELIABILITY AND VALIDITY

The term "norm-referenced measurement" is usually associated with test scores measured along a continuous, or more strictly, ordered scale and for which population "norms" are available. The subsequent use of these scores with individuals from the same population then allows a percentile ranking to be assigned to that individual. From the perspective of this chapter, however, this term is somewhat redundant. Since all the scores or estimates we have been discussing are population dependent they can all, in principle, have population norms derived for them. In the "mastery" case these norms will simply indicate the percentage in each category.

Most importantly, it is still necessary to decide the relevant reference population, and as we have indicated this choice is often difficult to make. With educational tests it is common to condition norms on age and also sometimes on gender. From the point of view of precision it may also be useful to condition, say, on social or educational background of parents or ethnic group. For certain purposes, for example to do with diagnosis, such precise conditioning may also be appropriate. Yet for other purposes such as selection, this would not generally be acceptable. Thus, there is generally no single set of norms for any one test or assessment which is equally suitable for all purposes. There is also the practical problem of actually constructing the norms using large representative samples, but discussion of this is beyond the scope of this chapter (see Levy and Goldstein, 1984).

We shall not dwell on the issue of "validity". Essentially there are two broad senses in which this is used. One is concerned with the extent to which the indicators used to construct a test actually do reflect the desired attribute, and the other is concerned with measuring the agreement between the test score and some external validating measurement (see Chapter 14). The former use of the term involves careful consideration of content, context and impact upon individuals. The most difficult aspect of the latter use of the term is the establishment of a suitable external criterion. If this can be found then we can use standard techniques for measuring association or correlation to provide a measure of validity.

The concept of reliability, however, is of central importance to the interpretation of test scores, whether these be continuous measurements or simple mastery judgements. For mastery or graded judgements it is common to formulate definitions of reliability in terms of the probabilities of missclassifying an individual as a "master" when in fact they are not, and vice versa. This raises no different conceptual issue from the continuous score case and we shall concentrate on the latter.

When we obtain a score for an individual this is viewed as an estimate of the underlying attribute and, in principle at least, if repeated independently under the same conditions would produce a distribution of scores. We may formalise this by writing the observed score for the ith individual as a function of a "true" underlying value

$$X_i = x_i + e_i \tag{4.5}$$

We can then define reliability in terms of a variance ratio as

$$R_X = 1 - \sigma^2{}_e/\sigma^2{}_X$$

which is a function of both the variance of the "measurement errors" $\sigma^2{}_e$ and the variance of the observed scores $\sigma^2{}_X$. Thus it is population dependent and in particular, if the measurement error variance remains constant, we can increase the reliability simply by referring the definition to a more heterogeneous population. This could occur, for example, if a reliability is estimated for students over a wide age range within which the mean score is changing and thus the variance across the whole age range is larger than the age-specific variance. The more fundamental parameter is the measurement error variance and this would normally be used when constructing complex statistical models. The importance of obtaining satisfactory estimates of these variances is that when a test score is used, for example as a predictor in a linear model, the presence of measurement errors can lead to inconsistent estimates of coefficients, etc. (see for example Goldstein, 1979).

Most of the standard procedures for obtaining estimates of reliability or measurement error variance are based upon identity link models of the form of equations (4.1) or (4.2) (Lord and Novick, 1968). These procedures can give lower bounds for these parameters, but unfortunately they rely upon the conditional independence assumption, and where this does not apply the lower bounds do not hold either (Goldstein and Wood, 1989). Ecob and Goldstein (1983) showed that using different assumptions it was possible to obtain very different estimates of these parameters in a large data set, and it would seem in general that there is a need for more empirical study of reliability estimation procedures.

CONCLUSIONS

We have not attempted to cover all the topics related to the aggregation and scaling of item test scores. Some of these are covered elsewhere in this volume (for example the issue of test bias in Chapter 6). We hope, however, that we have indicated the main issues so that the reader can be clear about the purposes and assumptions behind standard test scoring procedures and their strengths and limitations. Finally, we would like to make a few comments upon current test score construction and possible alternative models.

The techniques we have described view the individual who responds to a collection of test items as possessing an underlying attribute which we wish to measure. In this sense there is a clear analogy with the natural sciences where the assumption of an underlying attribute for many physical and biological measurements is a reasonable one. In the social sciences, however, the object of measurement is expected to interact with the measurement in a way that may alter the state of the individual in a non-trivial fashion. Thus, on answering a sequence of test questions about quadratic equations the individual may well become "better" at solving them so that the attribute changes during the course of the test. This is one way in which the conditional independence assumption can easily be violated.

It is not only standard psychometric models which make this assumption of a "static" attribute. More recent forms of "authentic" assessment (Madaus and Kellaghan, 1993) and those assessments associated with, for example, public examinations in the UK, also treat the individual as having a set of attributes and the task of the assessment is to capture or measure these as accurately as possible. In many situations, however, such a view is clearly either inefficient or misleading. In assessments where credit accumulation operates, for example, learning clearly takes place over time and the possibility of accumulating credits in different orders and retaking units is inconsistent with a static attribute model. In the UK such schemes, known sometimes as modular examinations, are becoming increasingly popular and they also operate where "course-work" or portfolios are used as part of an assessment. In such cases what is required is a dynamic model of assessment which takes into account the time ordering and any learning which takes place over the period of the assessment. There is information contained in not only the average over this period but also, for example, the rate of change. This would seem to be a fruitful area for the development of models for the aggregation of test items.

CHAPTER 5

Defining, Setting and Maintaining Standards in Curriculum-embedded Examinations: Judgemental and Statistical Approaches

Michael J. Cresswell

The Associated Examining Board, Guildford, Surrey, England

STANDARDS AND VALUES

This chapter analyses the problems of defining, setting and maintaining standards in curriculum-embedded public examinations. By recognising the setting of standards as a process of value judgement, the analysis presented here explains why successive recent attempts to set examination standards on the basis of explicit written criteria have failed and, indeed, were doomed to failure. The analysis also provides, for the first time, a coherent theoretical perspective which can be used to define comparable standards in quite different subjects or assessment domains. In addition, the general principles underlying the analysis apply equally well to other means and purposes of assessment, from competence-based performance assessments to multiple-choice standardized tests.

Values permeate assessment processes in any field and educational assessment is no exception. The values of the designers and operators of educational assessment systems influence every step of the assessment process. Deciding upon the broad domain of knowledge, skills and understanding which is to be assessed imposes particular values by identifying the attainments which are of interest. Selection, from that domain, of what is assessed in a particular instrument superimposes a second set of values concerning the relative importance of the different elements within the domain. In formal assessment systems, such judgements of relative importance are refined and codified by decisions as to the

Assessment: Problems, Developments and Statistical Issues.
Edited by H. Goldstein and T. Lewis.

number of marks or questions which will be used to assess each such element. Finally, judgement of the quality of pupils' performances brings to bear a further set of values.

The ways in which values influence the assessment process can be grouped into two broad categories: decisions concerning *what* is to be assessed and judgement of the *quality* of pupils' responses. To make the point in concrete terms, statements such as *the ability to recall historical fact will be assessed* and *each fact tested is of equal importance* are logically distinct from the statement *an acceptable performance is recall of 75% of the historical facts tested.* This distinction is key to an adequate analysis of the problems of defining examination standards. It is a central thesis of this chapter that much of the confusion surrounding examination standards is caused by a consistent failure to give adequate attention to both of these aspects.

This failure has its most visible effects in public argument about changes in educational standards over time. Typically, debate centres upon quantitative measures such as the proportion of an age cohort passing a particular examination at two different times. However, school curricula are essentially dynamic with teaching practices and curriculum content subject to continuous mutation to reflect current ideas and mores and to periodic major revisions introduced for theoretical or political reasons. In these circumstances, examinations separated in time must either measure different variables so as both to be valid tests of the different curricula being delivered on the different occasions when they are used, or at least one of them must be invalid on at least one occasion. In either case, direct comparison of quantitative measures based upon the two examinations does not provide valid objective information about changes in attainment over time.

The essential point here is perhaps best illustrated by an example. Euclidean geometry was taught extensively in British secondary schools up to the middle of the 1960s but little of it remains in the current (1995) mathematics curriculum. Consider two observers, one of whom believes that Euclidean geometry is an irreplaceable part of mathematics as a field of study and one who does not take this view. Comparisons of Mathematics examination results from the 1960s and 1990s will not enable these observers to reach agreement about changes in educational standards in mathematics since to do this they must be able to agree about the relative value of Euclidean geometry and what has replaced it. In the same way, defining examination standards must always involve a consideration of the value attached to the knowledge, skills and understandings which are assessed.

Thus, defining examination standards involves making value judgements concerning the attainment of those being assessed and, to have any meaning, these value judgements must address two aspects of the measured attainment: *what* attainments are assessed and the *quality* of the observed performance during which those attainments are demonstrated. In normal assessment practice, these two aspects are considered separately; *what* is determined when the examination syllabus is devised and is then operationalised within a particular examination on that syllabus whereas *quality* is judged once candidates have completed the examination. It is only this latter process with which the literature on setting standards in tests and examinations is normally concerned (see, for example, Berk, 1986). This is unfortunate because, although logically distinct, the two aspects *what* and *quality* interact with each other at a deep level. What is included in syllabuses and examinations is chosen to be appropriate given general expectations of candidates' performances and the quality of candidates' work can only be judged on the basis of the particular tasks upon which they are assessed.

The identification of examination standards with value judgements has a number of consequences. Firstly, value judgements generally have a bad name when it comes to reliability and consistency. This is particularly true of judgements about the *quality* of people's work where an uncomfortably close parallel is provided by critical judgements about aesthetic quality. Secondly, the value attached to *what* is assessed is likely to depend upon the purpose for which the assessment is made. Absolute examination standards are thus seen to be a chimera; standards can only be defined for a particular purpose and on the basis of the shared values of a particular community at a particular time.

The major consequences of this perspective on the problems of defining, setting and maintaining standards are explored below. The discussion is set mainly in the context of curriculum-embedded public examinations which provide qualifications for use in vocational or educational selection. However, reference is made, where appropriate, to relevant work on other types of assessment instrument. First, however, a brief technical description is given of current approaches to setting standards in two contexts: the British public examination system and multiple-choice testing.

AGGREGATION, AWARDING AND STANDARD SETTING PROCEDURES

Educational assessment normally involves summarising a large number of observations into a small number of indicators. In conventional standardised testing for example, pupils might tackle 50 questions and the number of questions answered correctly by each pupil will then be counted to provide a single total score. Similarly, in conventional public examinations, the marks from different questions within a paper are added together and then the marks from the different papers are themselves added to give a total mark for the examination as a whole. This process of combining observations of educational performance into a single indicator has come to be known as *aggregation*. It has two main technical functions. Firstly, it serves to enhance the reliability of the indicator because random errors in the individual observations tend to cancel each other out during the aggregation process. Secondly, aggregation enhances the validity of the indicator by enabling the assessment process as a whole to sample a variety of distinct facets of the domain of attainment being assessed. As part of this second function, conventional aggregation processes also permit differing weights to be explicitly given to different facets of the domain so as to define the nature of the attainment summarised by the indicator (Cresswell, 1987a).

Usually, the total scores which aggregation produces are subsequently converted on to a different scale which sets the pupils' results in a more general context. For standardised tests, this may be a norm-referenced scale with known mean and standard deviation. For some criterion-referenced tests, the reported scale might simply have two values: mastery and non-mastery. For most public examinations, candidates' results are reported in terms of a scale of letter grades which is common to all the examinations within a given system (for example, in Britain, General Certificate of Education (GCE) A-level examinations, which are the main means of selecting students for higher education, report in terms of five passing grades A, B, C, D, E and two failing grades: N and U). The process of converting pupils' public examination scores into grades is known as *awarding*. The parallel process of establishing mastery levels for conventional criterion-referenced tests is generally known as *standard-setting*. Both processes characteristically involve establishing one or more threshold scores on the total mark scale which define the grades corresponding to the intervening ranges of marks.

In British public examination practice, both for university entrance (GCE A level) and school leaving purposes (the General Certificate of Secondary Education (GCSE) examinations) this is done by a group of judges known as *awarders* who, by making judgements of the quality of candidates' work, identify the total mark which corresponds to each grade threshold. (The awarders include the *chief examiner* who also sets the question papers and co-ordinates their marking.) For the first examination on a new syllabus, the awarders can be said to be *setting* the standards for the grades; for subsequent examinations they can be said to be *maintaining* those standards. In both cases, however, broadly the same procedures are used. The awarders begin their task by reviewing the way in which the particular examination which they are awarding has performed on the particular occasion concerned. They receive an impressionistic report from the chief examiner about the way in which candidates have responded to the questions and, in particular, about any questions which proved unexpectedly difficult or easy and they review the statistics of the marks awarded, making comparisons with the papers used in previous examinations on the same syllabus. The awarders then scrutinise examples of candidates' examination work which have received different marks, judging the overall quality of each piece of work and the grade which it therefore merits. On the basis of these judgements of quality, the awarders make a tentative identification of the lowest mark which is associated with scripts they judge worthy of each grade. The procedure can be thought of as an application of the *contrasting groups* standard-setting methodology described by Livingston and Zieky (1982) in which candidates are assigned to the two groups by holistic judgement of the quality of their work. Scrutiny of scripts goes on until a consensus is reached about the threshold score which best distinguishes between the groups.

The awarders then consider statistical evidence in the form of the distribution of grades which would follow from using the marks which they have initially identified to partition the total mark scale into grades. This initial distribution is usually compared with the grade distributions from other examinations, but these comparisons are problematic because different candidates choose to enter for different combinations of school subjects. Moreover, even within a given subject, candidates choose to enter for one of a number of different examinations which are set by several autonomous GCE examining boards (or GCSE examining groups) on differing syllabuses (drawn up by the boards and groups) which adopt different approaches to the subject concerned. Systematic differences between the groups of candidates entered for different examinations are therefore to be expected. The comparison has greatest validity when standards are being maintained in a series of examinations on the same syllabus. Then, the comparison is with previous examinations in the series and the expectation is that there will not be major changes in outcomes from one year to the next for examinations where the number of candidates is large and the types of schools and colleges entering them are the same. Should there be large differences between the grade distributions from the two years, the awarders will reconsider some, or all, of their qualitative judgements by scrutinising further examples of candidates' work.

Having considered both the candidates' work and the statistical data, the awarders make a final composite judgement of the lowest mark which will be taken to merit the award of each grade for the particular examination concerned. These threshold marks for each grade are commonly called the *grade boundaries* and, as a set, define the ranges of total marks corresponding to the grades. More detailed descriptions of conventional British public examination grading procedures are given by Christie and Forrest (1981) and Good and Cresswell (1988). Both of these sources make clear the fundamental way in which the awarding process depends upon value judgements. Recently, the government body set up to regulate British public examinations (the School Curriculum and Assessment Authority (SCAA)) has published national specifications of awarding procedures for GCE and GCSE examinations (SCAA, 1994a, SCAA, 1995 respectively) which confirm that this is still the case.

In the United States of America, although interest in more authentic curriculum-embedded assessment has been growing in recent years, the historical prevalence of multiple-choice testing has meant that techniques for setting standards have focused mainly on the problems of determining *cutting scores* for such tests (but see Cross *et al.*, 1985, for an exception). Much American work has explored different methods of combining judgements about individual items to produce cutting scores for the test as a whole. Methods proposed by Nedelsky (1954), Angoff (1971) and Ebel (1972) have been the focus of many comparative studies (see, for example, Andrew and Hecht, 1976 or Mills, 1983). The Angoff method has been the most widely used and was given the highest rating in Berk's (1986) "consumer's guide" to standard setting methods. This method requires judges to say, for each item in the test, what proportion of a hypothetical group of minimally competent candidates would get it right. These proportions are then summed to provide the cutting score on the test as a whole for the level of competence concerned.

As in British public examination grade awarding procedures, work on standard setting in other contexts has highlighted the benefits of discussion among judges (Fitzpatrick, 1989) and of providing the judges with statistical data on the performances of candidates on the test in question (De Gruijter, 1985; Busch and Jaeger, 1990). In these and other ways (see, for example, Fehrmann *et al.*, 1991, on the benefits of training judges), much effort has been expended on improving standard setting techniques for multiple choice tests, but the subjective nature of the judgements involved has remained fundamental (Shepard, 1980) even if, as Rowley (1982) entertainingly illustrated and Cizek (1993) more recently argued, it has sometimes been obscured by the technology used.

Finally in this introductory section, it is worth mentioning some novel approaches to setting standards which were recently tried in England and Wales in connection with the assessment of the National Curriculum at ages 7, 11 and 14. Detailed descriptions of the approaches involved can be found in Hutchison and Schagen (1994). These new approaches used aggregation methods other than conventional addition of marks in an attempt to set standards in such a way that the results of the assessment would be as succinct as conventional examination grades but, none the less, would enable explicit descriptive inferences to be drawn about the nature of each pupil's attainment. This attempt was a failure and has since been abandoned in favour of more conventional aggregation and standard-setting procedures. Some of the theoretical reasons for this failure are implicit in the analysis given in this chapter but the new approaches are not explicitly considered here. Interested readers can consult Cresswell (1987b), Cresswell and Houston, (1991), Wolf (1993) and Cresswell (1994) for detailed critical analyses (see also Chapter 14).

STANDARD SETTING AS AN EVALUATIVE ACTIVITY

The reliability of standard setting procedures

If, as is argued here, standard setting is a subjective process akin to evaluating a work of art, immediate questions arise about its acceptability. Given the importance to individuals of the selection decisions which rest upon public examination results, the issue of subjectivity is clearly important. However, it is important to appreciate the sense in which *subjective* is being used here. The value judgements of awarders cannot be facts amenable to empirical verification or epistemological justification but they can, none the less, be the results of a rational process and be supported by reasons (Fogelin, 1967; Beardsley, 1981) if not pure deductive or inductive reasoning (Best, 1985). Value judgements based upon reasoned argument are not simply emotional or intuitive responses and should not, therefore, be assumed to be necessarily capricious or unreliable in the sense of being difficult to replicate.

In fact, there appears to be a tolerable degree of reliability in the value judgements made for public examination grade awarding purposes. Indirect general evidence for this comes in the form of the results of the many cross-moderation comparability studies which have been carried out over the years (Forrest and Shoesmith, 1985) and have implied that the value judgements of different groups of awarders agree reasonably well, most of the time. Little direct evidence of the reliability of awarding decisions is available. However, some was collected by Good and Cresswell (1988) who replicated the awarding meetings for their experimental examinations in French, History and Physics. Good and Cresswell (1988, p. 23) concluded that

"... different groups of grade awarders can reach decisions about final grade boundaries which are sufficiently similar to be acceptable, given the inherent imprecision of the examining process".

This conclusion was formed on the basis of the consequences for candidates' reported grades of the differences in judgement between parallel teams of awarders. In Good and Cresswell's study, the percentage of candidates whose subject grade changed if one awarding team's judgements were substituted for another's was 13% in French, 17% in Physics and 38% in History. Good and Cresswell point out that, although large, such changes are unexceptional in the context of the grade differences which occur between markers, even with very high levels of marking reliability. Work by Wilmut (1981) shows that, for the GCSE grade scale, 38% of candidates changing grade corresponds, approximately, to an inter-marker reliability coefficient of 0.96.

Much more work has been done on the reliability of the results of standard-setting methods in contexts other than public examinations. This has characteristically found larger differences between the results of different methods than between repeated applications of the same method (see, for example, Andrew and Hecht, 1976 or Cross *et al.*, 1984). None the less, agreement between individual judges using the same method has often tended to be poor (Brennan and Lockwood, 1980) and Angoff (1988) identified the reliability of standard setting methods as a matter in urgent need of attention.

Considerably better agreement has, however, been obtained among judges who have received training (Fehrmann *et al.*, 1991), had an opportunity to discuss their judgements (Fitzpatrick, 1989) or been given data on candidate performance (Busch and Jaeger, 1990). Morrison *et al.*, (1994) introduced statistical data and discussion into the Angoff method in a carefully controlled way and concluded that, particularly with suitably trained judges, the reliability of the resulting standards could be assured.

The use of explicit qualitative criteria

Nonetheless, more reliable awarding and standard setting procedures would be an advance and it appears reasonable to argue (as did Christie and Forrest, 1981) that reliability would be improved by the identification of the criteria which are used to make the judgements which determine candidates' grades. It is also argued (by sources as disparate as Nedelsky, 1954, and DES, 1982) that if these criteria could be identified, they could be written down and used to communicate the standards required for each grade to candidates, teachers and selectors. However, it is important to note the difference between the previous two sentences in the way in which they use the word *criteria*. Christie and Forrest conceptualise a grade criterion as an attribute to be assessed; the DES paper sees it as a standard. In the early 1980s, there was a considerable amount of work done in Britain on this ambiguous notion of *grade criteria* for public examinations (Hadfield, 1980, quoted in Christie and Forrest, 1981; Orr and Nuttall, 1983; SEC, 1984; Forrest and Orr, 1984; Orr and Forrest, 1984; Bardell *et al.*, 1984; SEC 1985; Long, 1985; SEC, 1986; 1987). Almost all of this work was based on the view that grade criteria should be standards; that is, as Nedelsky (1954) implied, they should be written statements which prescribe the level of attainment required to justify the award of a particular grade. It is worth noting that by 1984 Forrest and his co-workers also followed this line, explicitly confusing the two conceptions of criterion in the following definition: "[criterion-referencing is] relating the award of grades to specified *levels of attainment* in defined *aspects of the subject*" (italics added).

These various attempts to develop explicit standard-setting grade criteria concerned themselves almost exclusively with apparently observable qualities in candidates' work. This reflected an implicit underlying theoretical model which viewed the task of setting grade standards simply as the identification of appropriate qualities in candidates' work, rather than the formation of a value judgement based upon such qualities. The effects of this theoretical error became manifest in the work done by the examining boards in England and Wales at the request of the British government's advisory body of the time, the School Examinations Council (SEC, 1986; 1987). Although it was possible to write standard-setting grade criteria (either *ab initio* as in the original SEC work or as a result of perusing candidates' work), in use they proved not to apply to some examination performances which had, none the less, been awarded the grades in question by conventional procedures (Cresswell, 1987c). Nor was this the result of errors in awarding grades in the first place, since the examiners involved in the grade criteria studies characteristically avowed the appropriateness of the conventionally awarded grades. Similar results were obtained in the work carried out by Forrest and his co-workers (Forrest and Orr, 1984; Orr and Forrest, 1984; Bardell *et al.*, 1984) who, in the general conclusions published in the reports of all their studies, observed:

"performances in existing examinations that would result in the award of particular grades may not qualify for those grades if the criteria considered relevant were applied".

Thus, despite considerable efforts, no system of explicit standard-setting grade criteria for awarding public examinations has been made to work successfully, given that success means replacing holistic value judgements of quality with a set of explicit criteria. It is, of course, possible to challenge this definition of success, but the consequence of using explicit criteria which are not successful in these terms is that the examinations concerned do not meet commonly accepted requirements related to fairness (see Cresswell, 1987b; Cresswell and Houston, 1991 and Cresswell, 1994). In any case, the intention of the work on grade criteria was to make existing criteria explicit without changing the standards represented by the grades.

One reason for the failure of attempts to develop comprehensive explicit grade criteria is the multi-attribute nature of the attainments being assessed and the inability of the criteria to specify the weight which should be attached to each attribute when judging individual candidates' work. Wilmut and Rose (1989), who had been trying to use descriptive statements akin to grade criteria to set standards for the award of levels in a modular assessment scheme, summarised the position as follows:

"The real difficulty . . . has been the breadth of these descriptions; because they relate to a whole module they must encompass a wide range of attributes, and it is very difficult to characterise these adequately in statements which are brief enough to be readily usable for decision-making. There is, of course, no difficulty in summarising the descriptions in order to convey the 'flavour' of a level, but reduced statements of that kind are of little use for deciding about the worth of students' work. In practice, such work is often in several pieces, each of which has to be assessed separately, and is usually intended to exhibit several connected but separate attributes. Thus we may, for example, be attempting to assess a piece of project or fieldwork under the general objectives of:

- seek out information
- use that information
- communicate the results

Level descriptions have adequately to cope with these three objectives, properly indicating their relative importance, and be expressed in a way which allows us to make sensible decisions about pieces of work in which the objectives are met to different degrees. Thus, one student might be good at seeking out and using, but poor at communication, whereas another may use information less well, but get the communication about right. Both might merit a certain level, and the descriptions must enable us to see that this ought to be so."

As far as it goes, Wilmut and Rose's diagnosis of the problem is correct, but they still appear to believe that practical standard-setting criteria can be written which reduce the formation of value judgements to the simple identification of objective qualities. In fact, as the next section argues, the failure of every known attempt to construct and use such grade criteria is not a failure to do the job properly by those who have tried. It is an inevitable consequence of the fundamental nature of value judgements.

The nature of value judgements

That standard setting value judgements, particularly those made by groups of trained judges after discussion of all the relevant information, can often be reasonably reliable (see 'reliability of standard setting procedures' above) would not surprise authors such as Fogelin (1967) and Best (1985) who have made a study of the processes by which value judgements are formed. Although the process of evaluation is not one of deductive or inductive logic, it is none the less rational. Reasons can be adduced for evaluative judgements and these reasons can be assessed by well-accepted criteria. In particular, those reasons which are matters of fact can be verified and combinations of reasons can be assessed for consistency, one with another. Thus, although there is no way of *proving* the accuracy of value judgements, it is not true that rational debate cannot take place about them, nor that discussion cannot usefully clarify the reasons for such judgements and thereby persuade others of the accuracy of a particular judgement.

Nonetheless, it remains the case that two awarders might agree completely about the relevant reasons for their judgements but still differ in that judgement. This is because, understood in terms of deductive logic, value judgements can be seen as following from a set of premises (the agreed reasons) and at least one further prescriptive premise (Fogelin, 1967). In this way, Fogelin argues that value judgements are warranted prescriptions which recommend a course of action. Whether or not this is true of value judgements in general is a matter of debate (see Pole, 1961), but it appears true in the case of examination grades. The award of a Grade B, for example, amounts to the prescription to select this candidate in preference to one with a worse grade but not to select this candidate in preference to one with a Grade A; the warrant for this prescription is grounded in the candidate's measured attainment.

Clearly, the adoption of different prescriptive premises would be sufficient to explain irreconcilable disagreement between awarders who, none the less, shared the same view of the qualities of a candidate's work. Thus, if awarders are necessarily to agree about value, given that they agree about reasons, they must share the same prescriptive premise(s). Billington (1988) argues that prescriptive premises can also be discussed rationally, often in terms of the value of their consequences in other cases. However, there is a circularity here in the appeal to other value judgements to justify prescriptive premises and the consequence is that awarders are most likely to reach agreement after discussion of their respective views of candidates' work, if they share a common understanding about the prescriptive import of those qualities.

Sadler (1985, 1987, 1989) has written extensively about the way in which educational evaluation involves the use of *tacit standards* held by a *guild of professionals*. It is this aspect of the awarding process which enables awarders, at least in theory, to progress towards agreement about value judgements by discussing their reasons and prescriptive premises. It is noteworthy that, whereas Sadler's 1987 paper advocates the use of explicit written standards, supported by examples, to communicate the guild standards to others, his later 1989 paper recommends that the best way to teach students how to evaluate their own work is to give them direct, guided experience in evaluation which involves discussion of the criteria used in many specific instances.

This approach is also consistent with Sadler's other main contention: that the criteria which provide appropriate reasons for an educational value judgement differ according to the particular piece of work being evaluated. This has long been recognised as a feature of judgements of aesthetic value (see, for example, Aldrich, 1963). Indeed, the presence of the same quality may be reason to value one object highly but not another because of other features of the second object which change the value attached to the quality in question (Pole, 1961). Here is the fundamental reason why the application of concise sets of explicit written criteria does not replicate holistic value judgements made by suitably qualified judges and it dooms to failure the development of explicit grade criteria which define examination standards.

Note, however, that this is not to say that concise statements cannot be constructed which, in Wilmut and Rose's (1989) terms, "convey the flavour" of a grade. Such *grade descriptions* have been included in GCSE syllabuses, for example, since 1988. However, they describe a paradigmatic attainment worthy of the grade, rather than the attainment of every candidate awarded the grade and cannot, for the reasons discussed above, be used as criteria for judging the attainment of all candidates.

It is also worth noting that the theoretical analysis in this section is consistent with the normal practice of separating the function of scoring or marking candidates' work from that of standard setting or awarding. Although marking of assessment instruments other than multiple choice tests usually has an evaluative aspect, it is carried out by means of a mark scheme which specifies precisely which criteria are assessed and, through numerical mark values, indicates how strengths and weaknesses in different criteria are traded off against each other. This approach is usually justified in the interests of forming a reliable rank ordering of large numbers of candidates when many different examiners are, for practical reasons, involved in marking the candidates' scripts.

If marking were to be replaced with the direct judgement of every candidate's work in terms of grades, the same large number of examiners would all need to share the same tacit standards. For practical reasons, these examiners would also be denied access to the reasoned argument which, it was argued above, assists the formation of consistent value judgements. In addition, when responses to the same set of assessment tasks are being assessed, the use of a set of predetermined criteria with predetermined trade-offs is facilitated because the range of evaluatively relevant qualities is limited by the assessment tasks. (Note, in passing, that it is consistent with this analysis that assessment tasks such as essay writing, which admit of a wide range of responses, tend to be those where evaluation plays a larger part in the marking process.)

The range of possibly relevant assessment criteria and the tendency for them to be differentially relevant to evaluation is very much greater if responses to different assessment tasks must be comparably evaluated. This must be done, for example, if the grades from two different examinations on the same syllabus are to represent comparable standards or if the grades from examinations in different subjects are to be used interchangeably for selection purposes. The following section considers these aims and discusses some of the approaches to setting and maintaining standards which can be adopted to meet them.

DEFINING COMPARABLE EXAMINATION STANDARDS

The claim that a given grade represents the same standard of attainment, regardless of the subject or examination from which it comes, is a very strong one which lies at the heart of the use of public examinations as providers of information for selection purposes. Indeed, to argue the need for comparable examination grades is simply to restate the need for selections made on the basis of them to be fair in terms of the meritocratic philosophy which underpins their use (Broadfoot, 1986). What constitute meritocratically fair selections? There is a strong strand in meritocracy, borne out by the behaviour of selectors, to the effect that individuals should be selected on the basis of their current attainment. Since selections are made from among candidates who have taken different examinations, this has led to a perceived need to establish quantitatively equivalent levels of attainment across qualitatively different assessment domains. This is formally impossible in terms of the theory of educational measurement (see, for example, Goldstein, 1986b). Comparability is not, however, a purely technical matter and can only be fully understood in relation to the social purposes for which examination grades are provided. Below, in the section on "socially defined standards", a new definition of comparable standards is proposed which gives appropriate emphasis to the social function of public examinations.

Because of the manifest importance of comparability for the fairness of selection, there have been many studies attempting to establish whether or not the grades from particular public examinations reflect comparable standards. The GCE examining boards in England and Wales published two summaries of the 51 such studies which had been carried out up to 1985 (Bardell *et al.*, 1978; Forrest and Shoesmith, 1985). The Schools Council published a series of major reports in the 1970s (see Willmott, 1980) and since 1985 a further 22 comparability studies have been carried out for GCSE and A level examinations with the reports being published by, and available from, the examining boards and groups. These studies cannot be reviewed in detail here, but the methodologies which have been adopted are of interest. They can be divided into two main types: those which involved statistical analysis of the distributions of candidates' grades (or equivalent analyses) and those (called cross-moderation studies) which depended upon examiners' judgements to identify comparable work from the examinations being studied. Some of the studies used both approaches.

Although rarely explicit about their theoretical perspective, those responsible for comparability studies which analyse the distributions of candidates' grades seem generally to subscribe to the erroneous view that comparability is a technical problem with a technical solution within measurement theory. The studies using examiners' judgements have also been largely atheoretical, simply taking it as reasonable that examiners are able to judge comparable standards of attainment on different assessment domains. Where they have an identifiable theoretical base, however, it appears to be to treat examiners' judgements as a criterion variable, assuming that scripts awarded the same average judgement should, on average, be awarded the same grade in comparable examinations. As a result, some such studies (for example, Houston, 1980) have gone to considerable lengths to obtain agreement among the examiners involved about the judgemental criteria which they use. This conception of the cross-moderation methodology also identifies comparability as a purely technical problem of measurement theory.

Since GCSE examinations were introduced in 1988, some subjects have been examined using differentiated papers. In these examinations, candidates taking different combinations of papers, which are designed to differ in difficulty, are awarded a grade on a common scale. The problems of setting comparable standards on the different combinations of papers used in such examinations were studied in depth by Good and Cresswell (1988).

As far as comparability of standards in contexts other than public examinations is concerned, the issue seems to have engendered much less interest. This appears to be because interchangeability of results from a wide range of different assessments has rarely been an issue. Within a particular series of tests, the equating of scores (Holland and Rubin, 1982) from successive forms has generally been seen as a sufficient technique for maintaining standards (De Gruijter, 1985). This approach, however, is not as unproblematic as it may seem. Two technical requirements which must be met are that identifiable subpopulations must not have different equating functions and, if a reference instrument is used to equate tests indirectly, it must relate in an identical way to the tests being equated. These requirements are discussed later in this section.

In any case, without strictly parallel forms, test equating is formally impossible, although a more relaxed requirement of score *equivalence* might be sufficient for the purpose of maintaining standards (Goldstein, 1986b); see also Chapter 4. Moreover, there are sometimes practical reasons, such as concern about the security of test forms, which make equating studies difficult to carry out. There may also be problems with the adequacy of the data used for equating, particularly if cut scores need to be set at fairly extreme points on the score scale. For these reasons, Norcini (1990) used the Angoff method to set standards afresh on successive forms of a test. He concluded that this approach, with a correction for observed differences in the lenience/severity of different groups of judges, was technically preferable to score equating. If the score equating approach is not used, all the issues discussed in this section are as relevant to multiple choice tests as they are to more authentic curriculum-embedded assessment techniques.

Comparable standards defined statistically

What has *comparable*, as applied to examination grade standards, normally been taken to mean in previous statistically based studies? In general, it has not been taken to mean that an *individual* taking two comparable examinations should necessarily be awarded the same grade. Not only are the assessments involved subject to various sources of error which would prevent this outcome from routinely occurring, but it is also accepted that individuals attain differently in different assessment domains. Thus, the notion of comparability applied to public examination results is concerned with groups of candidates. For example, talking about the particular case of comparability between examining boards (and assuming a great deal about other variables relevant to performance in public examinations which will be considered shortly) the Forum on Comparability set up by the Schools Council (which had a general responsibility for British public examinations until the mid-1980s) said:

"... the expectation is that had a group of examinees followed another board's syllabus and taken its examination, they might reasonably be expected to have obtained the same average grade." (Schools Council, 1979).

The definition of comparability implicit in this quotation can usefully be elaborated to cover the comparability of every grade, by replacing the reference to the "average grade" by a reference to the *distribution* of grades for the group of candidates. On this basis, a variety of definitions of comparable examination standards can be explored by considering some of the different ways in which differences between grade distributions have been analysed in previous studies.

Clearly, for analysis of examination grade distributions to provide a reliable indication of the comparability of grade standards, the grade distributions need to depend only upon those standards. In fact, of course, grade distributions also reflect the attainments of the candidates who take the examinations and these attainments are the result of the interaction of many different variables. As noted earlier, systematic differences between the self-selected groups of candidates who take different British public examinations are to be expected and the various statistical techniques which have been used in Britain for investigating examination comparability by analysing grade distributions attempt to control for differences in the attainment of the candidates taking the examinations in one of two ways. Either an independent measure of attainment is employed or indirect control of attainment is attempted through school and student variables which influence it. Once this has been done, it is argued, any remaining differences between grade distributions indicate lack of comparability between the examinations.

Three major issues immediately arise, however. First, candidate attainment is, itself, affected by features of the examinations and their syllabuses (examination variables) as well as characteristics of the candidates' schools (school variables) and characteristics of the candidates as individuals (student variables). As a result, controlling for differences between different groups of candidates in terms of attainment will, in passing but unavoidably, also control to some extent for the relevant features of the examinations and their syllabuses. These features include the demands of the skills and subject content specified in the syllabus and its organisational aspects. Syllabuses can differ in both these respects and both may affect the motivation of pupils and the ease with which they are taught and learn. For example, at one extreme, a syllabus can be presented as a simple list of skills and content; at the other extreme, the same skills and content can be usefully structured and accompanied by examples and guidance for teachers. The question of importance in the present context is whether a syllabus (and associated examination) which makes the subject more accessible or more motivating and thus produces a grade distribution which is skewed towards the better grades, therefore effectively sets lower standards for the award of grades than an obscure syllabus which makes learning difficult. Questions of this type must be answered before any statistical analysis of grade comparability can be interpreted.

The key to the answer lies in the function of the examinations. If this is the provision of information for meritocratically fair selection processes in which *current attainment* is the selection criterion, then fairness requires that those who study syllabuses which organisationally facilitate learning should have a greater chance of being selected than those who study less enabling syllabuses. It follows that they should, in their examination grades, be rewarded for their higher attainment even though it is a consequence of the organisation of the particular syllabus which they have followed. The corollary of this is that different grade distributions from two examinations which differ only in the organisational aspects of their syllabuses do not necessarily indicate a lack of comparability in the grading standards being applied.

On the other hand, if two syllabuses are judged to be of equal value but none the less differ slightly in terms of the intellectual demand of their content, consequential differences between their grade distributions would imply a lack of comparability because it is meritocratically unfair for a candidate to have less chance of being selected because he or she showed less attainment of a harder assessment domain. (Indeed, it is concern about differences of this sort which motivates the demand for comparable grade standards in the first place.) However, differences in syllabus demand are also likely to affect the motivation of pupils and differentially to facilitate teaching, and it was argued in the previous paragraph that motivational differences between syllabuses are a legitimate reason for differences in their grade distributions.

Because of interactions of this type, in practice it is impossible to distinguish between the effects of differences in the intellectual demands of syllabuses and the effects of differences between their organisational aspects. As Goldstein (1986b) points out, comparability studies based upon the statistical analysis of grade distributions are forced to ignore such differences and generally take as a working assumption that the effects of syllabuses and examinations upon teaching, learning, examination entry policy and so on are identical for all the examinations studied.

The second theoretical problem with definitions of comparable standards which depend only upon identical grade distributions, concerns identifiable subgroups of candidates. Even when two grade distributions are identical for the whole groups of pupils whose attainment they describe, they are frequently not identical for well-defined subgroups within those groups. For example, the differences between boys' and girls' performances in GCSE examinations are well known and depend, at least in part, on the assessment techniques used (see, for example, Murphy, 1982; Cresswell, 1991). Thus, if two examinations using different techniques have identical grade distributions for boys and girls combined, they will not necessarily have identical grade distributions for the boys and girls considered separately. It is unclear how this can be accommodated if the test of comparable standards is identity of score distributions, since the meaning of comparability is unclear if it is limited only to a particular subgroup of candidates. Certainly, comparable standards defined only for a particular subgroup of candidates are insufficient to permit the general use of examination results in selection. Only if the control of the other variables which determine grade distributions also removes any observed differences between well-defined subgroups of candidates, can the difficulties raised in this paragraph be avoided.

The third, and most fundamental, problem with purely statistical methods of studying comparability concerns the content of the syllabuses. Christie and Forrest (1981) pointed out that examination standards can only be comparable if the syllabuses concerned define assessment domains which are appropriate to the "particular subject at a particular level of education". That is to say, in the terms of the opening section of this chapter, the *value of what is assessed* must be comparable, as well as the grade distributions. As a thought experiment to illustrate this point, consider a GCE A level Physics syllabus which was entirely descriptive and did not contain any mathematical treatment of the phenomena covered. An examination could undoubtedly be set on such a syllabus which would produce the same grade distribution for a given group of candidates as they achieved on a more conventional Physics syllabus. Would, therefore, the grade standards of the two examinations be comparable? The answer to this question clearly depends upon the value given to a non-mathematical Physics syllabus at A level. Thus, defining comparability of examination grading standards solely in terms of identical grade distributions is inadequate unless it can be assumed that the syllabuses upon which the examinations are based define assessment domains of equal value. This assumption is therefore crucial to all statistical approaches to setting comparable examination standards but, for the sake of the discussion, it will not be challenged immediately. In the section "Comparable standards defined socially" below, however, the vital importance of the valuation of *what* is assessed is reintroduced and a new theoretical basis for the definition, setting and maintenance of standards is proposed which accommodates it.

As noted earlier, a striking feature of most of the statistical work on examination comparability which has been carried out is the lack of an explicit theoretical basis. In particular, few published statistical studies, from the Schools

Council work in the 1970s (Willmott, 1980) to the recent work by Fitz-Gibbon and Vincent (1994), have attempted to test the three crucial assumptions set out above: that different syllabuses have identical effects upon motivation, teaching, learning, school entry policy and so on; that the observed relationships between the grade distributions are the same for identifiable subgroups of candidates; and that the value of what is assessed in the examinations studied is comparable. Instead, those responsible for comparability studies which analyse the distributions of candidates' grades seem, relatively uncritically, to have adopted one of the variety of implicit definitions of it which are explored below.

(a) The no-nonsense definition

Two examinations have comparable standards if the distributions of grades which they produce are identical.

From the foregoing discussion, it will be clear that this definition is inadequate. It assumes that in terms of relevant student variables (such as ability, prior achievement, or motivation) and relevant school variables (such as general effects, effectiveness of subject teaching, or entry policy) together with the effects upon these of the two syllabuses and examinations, the groups of candidates entering the two examinations are identical. Most of these assumptions are known not to hold in practice. In addition, comparability defined this way will not always hold for subgroups of candidates considered separately even if it does for the group as a whole. None the less, the no-nonsense definition of comparability is worth mentioning because it is frequently used in press discussion of examination results. It has also been used by British government advisory bodies as the basis for querying, with the examining boards and groups, the comparability of particular examinations and has been used during the deliberations of the Independent Appeals Authority for School Examinations (see, for example, IAASE, 1993) to which candidates can appeal if they believe their public examination results to be in error.

(b) The same-candidates definition

Two examinations have comparable standards if, when the same group of candidates is entered for them both, the distributions of grades which they produce are identical.

This definition assumes that motivation, prior achievement and the influence of relevant school variables, together with the effects upon these of the two syllabuses and examinations, are identical when the same candidates tackle two different syllabuses and examinations. There is no reason why these assumptions should hold in practice simply because the same students are involved. None the less, the intuitive appeal of this definition of comparable standards is considerable; it is used by teachers who enter candidates for examinations in the same subject with different boards and expect to get identically distributed results. This definition is also the basis for a particular approach to comparability between examinations in different subjects, known as *subject pairs analysis*, which has long been controversial on account of the assumptions just outlined (see Forrest and Shoesmith, 1985).

It is worth briefly mentioning one interesting condition under which the assumptions of the same-candidates definition might be thought of as axioms, rather than challengeable assumptions. If all the pupils in a particular age cohort take two examinations in different subjects, then it appears reasonable to define comparable standards between the two subjects concerned in terms of identical grade distributions from the two examinations. This situation is most closely approached in the British public examination system by GCSE examinations in English and Mathematics. The argument is as follows: it is difficult to see what meaning can be attached to the notion that the sum total of all the English teaching in the educational system is more or less effective than the sum total of all the mathematics teaching. Similarly, if there is, on average across all those who study them, a motivational difference between the two subjects, is this not most parsimoniously treated as simply characteristic of learning in the two subjects?

Fundamentally the same argument can be applied to all the variables upon which the attainment of the pupils in the two subjects depends. If it is accepted, it follows that, on condition that the examinations are taken by the entire age cohort, the same-candidates definition of comparable standards is theoretically coherent and might be useful. However, the implications of differential comparability for different candidate subgroups would need to be dealt with if it occurred and there would remain a significant problem if the relationship between performances in the two subjects changed over time. In these circumstances, either comparability between subjects or across time could be maintained, but not both. Note that although the age cohort condition is not generally met in British public examining, it is quite closely met by the statutory assessments at other ages of the National Curriculum in England and Wales.

(c) The value-added definition

Two examinations have comparable standards if two groups of candidates with the same distributions of ability and prior achievement receive grades which are identically distributed after studying their respective syllabuses and taking their examinations.

This definition assumes that, in terms of other relevant student variables such as motivation and relevant school variables together with the effects upon these of the two syllabuses and examinations, the two groups of candidates are identical. Most of these assumptions are known not to hold in practice, so analyses which simply allow for ability and prior achievement are unlikely to give reliable information about comparability. However, if the prior achievement in question is the result of schooling in the institutions which enter the candidates, allowing for it may also partially, but to an unknown extent, allow for the effects of some school variables. This definition has frequently been used; most recently by Tymms and Fitz-Gibbon (1990), Fitz-Gibbon and Vincent (1994) and Tymms and Vincent (1995), but most extensively by the Schools Council researchers in the 1970s (Willmott, 1980) who sometimes used a specially written reference ability test, known as *Test 100*, as a statistical control.

Because this definition has been extensively used in the past and is now enjoying something of a comeback, it is worth considering its underlying rationale from another angle. In particular, although Fitz-Gibbon, Tymms and Vincent do not discuss the theoretical basis of their studies at all, there does exist a theoretical perspective which appears to make the assumptions listed in the previous paragraph unnecessary. This perspective starts from the premise that it is ability or aptitude which should be the basis for selection, rather than attainment, and treats differences in measured attainment from subject-specific examinations as error caused by the vagaries of schooling and the examining process. Differences in measured comparability between candidate subgroups is seen as evidence of bias in the assessments. There is some evidence that the thinking behind the Schools Council reference test comparability studies followed this line; Nuttall and Willmott in 1972 suggested that there was a case worth considering for using "a single general intelligence test" in place of public examinations for selection purposes.

However, this quotation neatly illustrates the theoretical incompatibility between the use of public examinations of attainment for selection purposes and the use of ability measures to investigate their comparability. The only valid theoretical justification for using ability measures to establish examination comparability, automatically implies that the examinations are inappropriate tools for the selective purpose. On the other hand, if examinations of attainment are appropriate for selective purposes, then the use of ability measures to study comparability is inappropriate. It was noted earlier that the meritocratic philosophy which underpins the use for selection purposes of assessments made by the education system is more consistent with those selections being made on the basis of attainment than of aptitude. Moreover, in the light of Wood's (1986 and 1991) trenchant criticisms of aptitude testing, this seems just as well. Clearly, if the use of attainment as a selection criterion is accepted, any reference tests which are used to study comparability must be measures of current subject attainment. This brings its own problems which are set out in the next section.

(d) The equal-attainment definition

Two examinations have comparable standards if, for two groups of candidates with the same distributions of attainment, they produce grades which are identically distributed.

This definition is perhaps the one which naive observers would give if asked to define comparable grade standards in distributional terms. It is a direct application of the principles of fair meritocratic selection based upon current attainment. It avoids the difficulties caused by the large number of variables which affect attainment by controlling directly for that attainment. However, this definition is essentially circular and assumes a solution for the very problem of assessing different attainments on the same scale which it is intended to solve. It requires comparability first to be established between whatever instrument is used to assess the current attainment of the candidates and each of the examinations being studied before the comparability of the examinations themselves can be investigated. How is this to be done, other than by a further independent assessment of the candidates' attainment? This question makes the infinite regress involved in this definition obvious. The definition was, none the less, the basis of some of the Schools Council comparability studies of the 1970s which used attainment tests as reference instruments. However, these studies were discontinued because of the impossibility of constructing such a test which could be independently shown to be equally relevant to the differing assessment domains of the examinations being studied (Forrest and Shoesmith, 1985). Newbould and Massey (1979) gave a convincing demonstration that exactly the same problems arise if the reference instrument is a common element of the examinations themselves.

As noted earlier, some cross-moderation studies have also implicitly adopted this definition. Although they do not involve formal comparisons of grade distributions, some such studies have attempted to define a common criterion variable for use by the examiners when they are judging the quality of scripts from the examinations being studied. Comparable grade standards are then defined as the award of the same grade, on average, to candidates judged to have the same attainment on the common scale. This is effectively the equal-achievement definition given above. This approach is clearly subject to exactly the same theoretical problems as the use of a reference test. Christie and Forrest (1981) demonstrated the practical consequences of these by re-analysing Houston's (1980) data and concluding (p. 6) that:

> "... no conclusions should be drawn on the basis of an equally weighted composite [of three agreed criterion variables] for the simple reason that each board differs in the emphasis it accords each criterion ...

(e) The similar-schools definition

> Two examinations have comparable standards if two groups of candidates who attend similar schools receive grades which are identically distributed after studying their respective syllabuses and taking their examinations.

This definition assumes that, on average, "similar schools" are identical in terms of the school variables relevant to candidate achievement (for example, general effects, effectiveness of subject teaching and entry policy). The two groups of candidates are also assumed to be identical in terms of relevant student variables (such as prior achievement, or motivation) together with the effects upon these of the two syllabuses and examinations. Clearly, much hinges upon how similar the similar schools are and the extent to which controlling for schools also controls for relevant student variables. Delta analysis (Quinlan, 1993) is an analytic technique based upon this definition which has recently been used by the examining boards and groups in some of their comparability studies. However, with the data presently routinely collected about the schools and colleges which enter candidates for public examinations, large differences are known to exist within similar schools. Although this is a problem of current practice, rather than theory, the implicit control of student variables using the school variables as surrogates is an unavoidable limitation of this definition.

(f) The catch-all definition

Two examinations have comparable standards if two groups of candidates with the same distributions of ability and prior achievement who attend similar schools with identical entry policies, are taught by equally competent teachers and are equally motivated, receive grades which are identically distributed after studying their respective syllabuses and taking their examinations.

This definition is included here, not because it has been used in any published public examination comparability study, but to raise the question of why it has not been used. Although it remains prey to the theoretical difficulties of any definition of comparability which depends upon the identity of grade distributions because it, perforce, assumes that the effects upon the student and school variables of the two syllabuses and examinations are identical, this definition is the logical extension of previous statistical work on comparability. The collection of explanatory data about schools and students which the use of this definition would involve has been done in many other studies (for example, Brimer *et al.*, 1978; Cresswell and Gubb, 1987), would not be difficult in practical terms and would enable the issue of subgroup comparability to be explored. The only comparability work which seems to have been done along these lines is that begun by Nuttall and Armitage (1984), but similar work on the comparability of public examination standards would be well worth while and could make use of recent advances in multilevel modelling techniques (see Goldstein, 1995).

Comparable standards defined socially in terms of value judgements

The foregoing review of statistical approaches to the study of examination grade comparability reveals the existence of several different methodologies, each with its own implicit definition of comparable standards. However, none of these methodologies and definitions is satisfactory within measurement theory and the need for comparability studies to assume that meaning can be given to quantitiative comparisons between qualitatively differing attainments has often caused concern. For example, Johnson and Cohen (1983) say:

"A situation can be envisaged where this diversity [between assessment domains] is so great that there would be no consensus among subject specialists that the schemes concerned were equally valid under the same subject title; in such cases the issue of comparability becomes meaningless."

Most authors (for example, Forrest and Shoesmith, 1985) have agreed with Johnson and Cohen that the continuing utility of the public examination system as a provider of information for selection purposes depends upon the possibility of making sufficiently good *approximations* to valid comparisons between attainments in differing domains. A pragmatic case can be made for this point of view when the issue is comparability between different examinations in the same subject, but it leaves as meaningless the notion of comparability across different subjects. One of the major consequences of this was the cessation of any large-scale published British work on comparability between examinations in different subjects from the mid-1970s until Fitz-Gibbon and Tymm's (1994) study which adopted the value-added definition. None the less, the examining boards and groups have continued to claim, if only by implication, that the same grade represents the same standard of attainment in any subject and, in general, selectors have continued to behave accordingly. Recently, the British legal requirement of publication of public examination results as indicators of the success of individual schools has given a new importance to defining comparable standards across subjects because, otherwise, different mixes of subjects taken by different schools will affect the rankings of those schools in the published tables.

In this section, a new definition of comparable standards is proposed which, it is argued, provides an explanation for the long-term success of the public examination system as a provider of information for selection purposes, despite the apparent theoretical impossibility upon which it is based. The new definition also provides a theoretically coherent meaning for the notion of comparability between examinations in different subjects and accommodates the need to consider the relative value of *what* is assessed in different examinations.

Public examination grades can be likened to a currency with which candidates buy entry into education or employment. Developing this analogy a little further, the intrinsic value of banknotes does not match their face value, but commerce functions because it is commonly agreed to accept them at face value. (This agreement is greatly strengthened, but not guaranteed, by the underwriting of a national bank.) Similarly, educational and vocational selection processes can proceed provided that there is common consent that comparability exists between the grades from different public examinations. That is to say, provided that it is accepted that a given grade from one examination represents attainment of equal value to the attainment which earns the same grade in other examinations.

From this starting-point, it is possible to define comparable standards in social terms involving human judgements of value, as follows:

> Two examinations have comparable standards if candidates for one of them receive the same grades as candidates for the other whose assessed attainments are accorded equivalent value by awarders accepted as competent to make such judgements by all interested certificate users.

There are several important points to note about this definition of comparable standards. First and foremost, it avoids the formal impossibility inherent in the notion of quantitative equivalences between qualitatively differing attainments which undermine definitions rooted in the theory of educational measurement. This is because the new definition does not require such attainments to be quantitatively equivalent. Instead, by taking seriously the identification of awarding as an evaluative process which was made in the introductory section, the new definition requires equivalence to relate only to the *value* given to those attainments by the awarders. The nature of value judgements has long been an area of considerable philosophical debate (see, for example, Pole, 1961; Bambrough, 1979; Best, 1985). However, one issue in that debate is particularly significant in the context of examination standards. This is whether value judgements ascribe a property (or properties) to the objects being judged. As far as judgements of educational attainments are concerned, French *et al.*, (1987) maintain that they do not. Among many others, Ayer (1946), Fogelin (1967) and Billington (1988) have argued the same case in general, albeit from quite different perspectives. Under the proposed definition of comparable standards, examination grades, by reporting value judgements, therefore report human responses to the pupils' measured attainment, rather than the attainment itself. From this perspective, there is no *formal* impediment in the way of assigning equal value to dissimilar attainments.

The second point to note about the proposed definition of comparable standards is the price which has been paid for this belated philosophical justification of the notion that they can be defined across differing assessment domains. This price is the need to accept that there is no external and objective reality underpinning the comparability of results from different examinations. However, the subjectivity which is involved is no more and no less than that which, it has already been argued, is an inherent part of all standard-setting methods. As the section on the reliability of standard-setting procedures makes clear, such subjectivity need not necessarily lead to capricious or unreliable (in the sense of being difficult to replicate) judgements.

The third point to note is the *relativistic* aspect of the proposed definition of comparable standards. It refers to the *acceptance* of *all interested certificate users* and, by doing so, implies that other users in other times and places might dissent from the awarders' evaluations of pupils' attainment. Comparable standards as defined here can, therefore, only be established in a particular social context and, for users who do not accept the awarders' competence, will not be achieved. Clearly, the larger the group of examination users who are prepared to accept the awarders' competence to make the required value judgements, the more useful are the examination certificates. The group of users prepared to accept the evaluations of the awarders must include most candidates, parents, teachers and selectors if the examination system is to fulfil its purpose effectively.

This is not to say that, given the opportunity, all users would necessarily make value judgements identical to those of the awarders; indeed, they need not agree with the awarders' evaluations at all, as long as they agree to abide by them. In practice, however, acceptance is only likely to be forthcoming from any particular user on a continuing basis as long as the awarders' evaluations differ only to some small extent from their own judgements, however informal or uninformed the latter may be. Those setting grade standards must therefore either attempt to represent the views of most users or persuade the users that the standards which they set are comparable. The importance which the British examining boards have always attached to studying and reporting upon comparability can be seen as an implicit recognition of these requirements. It also demonstrates expertise in the techniques of educational assessment and thereby strengthens the claim of competence to set standards which is made by the examining boards.

It seems likely that acceptance by users will be more likely if the procedures used to set the standards are transparent and public knowledge. Cizek (1993) has written persuasively that the legal notion of *due process* can be used to underpin standard setting procedures. In Britain, the recent introduction of procedural codes of practice governing public examinations (SCAA 1994a and 1995, the latter having legal force) are a move in this direction which should help to maintain user acceptance. Indeed, this was the explicit reason for their introduction. It seems probable that many users are prepared to accept the standards set by examining boards because they do not believe themselves competent to make the required evaluative judgements for themselves. However, returning to the currency analogy with which this section began, the claim of competence to set accurate standards using appropriate procedures which is made by examining boards parallels the underwriting of national banks and is theoretically vulnerable, in just the same way and with similarly catastrophic effects, to a run on confidence.

In practice, however, the system appears to be reasonably robust and *general* acceptance continues of the competence of examining boards to award comparable examination grades, despite clear historical instances where some users did not accept their judgements. This was the case, particularly, for the old British CSE and GCE O-level examinations. The examining boards' judgements, based upon their evaluations as supposedly competent judges, that CSE Grade 1 was comparable to an O-level pass were accepted by most teachers and selectors in education, but never fully accepted by selectors in the vocational area. Whatever their own view, some pupils and parents, concerned about subsequent vocational selection processes, therefore behaved as if they, too, did not accept the examining boards' judgement about the comparability of the two systems. The failure of sufficient certificate users to accept the evaluations of the awarders in this case was partly a result of, and partly a reason for, the failure of CSE examinations ever to achieve parity of esteem with O levels. This led to the introduction of the GCSE and it may not be unconnected that the GCE boards, which operated the more highly esteemed O levels, have come to dominate the examining groups which now offer GCSE examinations.

In Britain at the time of writing (early 1995), there is considerable effort being put into achieving parity of esteem between GCE A-level examinations and General National Vocational Qualifications (GNVQs). Whether this is successful will depend upon the reactions of the users of the qualifications, but the CSE experience suggests that expedients such as the recent Government renaming of Level 3 GNVQs as "Vocational A levels" are unlikely to be sufficient. It is unknown to what extent the *general* claim of competence by the examining bodies involved will be damaged if users reject the claimed equivalence between these two particular examination systems (see Chapter 14).

The fourth important point to note concerns empirical tests of the comparability of standards. The proposed social definition of comparable standards permits comparability to be tested in a theoretically sound way. If comparability is defined in terms of the evaluations of awarders who must be accepted by users as competent, there are two empirical questions to ask: are the awarders' evaluations competent and are they accepted as competent? Taking the second question first, it would be difficult in practice but theoretically possible to ask any defined group of certificate users if they accepted the awarders' competence to make the evaluations of attainment implicit in the grades awarded from particular examinations. A less direct test of acceptance would be whether or not the users accepted the examinations concerned as fair. This test exploits the essential reason why comparable standards are important: the need for selection decisions to be meritocratically fair. Indeed, users' views on the fairness of the examination system as a whole could be tested, although they have not been, and there would be significant problems of definition to overcome about what represented a sufficient degree of user acceptance to support a general claim about comparable standards throughout the system.

Turning to the issue of the awarders' competence itself, it is a prerequisite for comparability between specific examinations that the judgements made by the awarders of one examination agree with the judgements of the awarders of a supposedly comparable one. Here the traditional cross-moderation approach to the study of comparability is appropriate and now has a clear theoretical basis although, from the perspective of the social definition of comparability, cross-moderation studies are concerned with conditions which are necessary but not sufficient to establish comparability. However, under the new definition, comparability between any examinations (including those in different subjects) can, in theory, be usefully studied using cross-moderation methodology. The practical problems of doing so in different subjects would certainly include finding awarders sufficiently knowledgeable to make the required value judgements in more than one subject. The nature of the knowledge required to make awarding judgements was discussed in detail in the section on the reliability of standard-setting procedures.

The final, and possibly most important, point to make about the proposed definition of comparable examination standards is that it also deals with the issue of the value of the syllabuses upon which they are based; that is, *what* is assessed. In the section on statistically defined comparability, it was argued that, unless the assumption that syllabuses are of equal value holds, statistical approaches to defining examination standards do so only in a very narrow technical sense which does not adequately reflect the normal meaning of the term *comparable*. Such a technical definition of comparable standards could imply, for example, that a high typing speed represents a standard of attainment comparable to a postgraduate degree in English. There is not, of course, any *objective* basis for deciding whether or not these attainments, so different in nature, are comparable in standard. However, to assert that they are comparable is not consistent with the value which is currently given to them in British society as measured by the rewards in terms of pay and social status which are given to those holding the jobs for which they act as qualifications (for example, typists and university teachers). Under the proposed social definition of comparable standards, awarders can, and must, take into account the wider social value of the syllabuses followed by the candidates if they are to make judgements of the value of candidates' attainments which are accepted by users as appropriate for the selective purpose. The necessary and sufficient test of the success of the awarders' attempts to do this is whether or not such acceptance is forthcoming.

SETTING OR MAINTAINING STANDARDS?

In normal public examining practice, an awarding meeting is held to set standards on every successive examination on a particular syllabus. It is not, however, made explicit by the examining boards whether the function of these meetings is the successive application of a particular set of independent evaluative standards or simply the application of the grading standards used in the immediately preceding examination to the current examination. In the second section of this chapter these approaches were called the *setting* and *maintenance* of standards, respectively. Of course, in theory the same result should be produced by either setting or maintenance, assuming that the evaluative process works perfectly every time.

Maintaining standards

However, the distinction does have considerable significance. Despite all that has been said in this chapter about standard setting as an evaluative process and about definitions of comparable standards, if the focus is *only* upon the maintenance of standards between successive examinations on an unchanging syllabus, some statistical approaches to comparable awarding become more viable. In particular, the same-candidates definition of comparability offers a theoretically coherent approach. All the assumptions of that approach are reasonable if the same group of candidates, having reached the end of their course, sit two examinations on the same syllabus. Thus, it would be theoretically possible to adopt the same-candidates definition for the maintenance of comparable standards between successive examinations on an unchanging syllabus. Despite certain practical difficulties, the possibility of maintaining British public examination standards in this way would be worth exploring. As noted in the section on comparability definitions, this approach is the one usually recommended in the literature about setting standards on other types of assessment instrument (see, for example, De Gruijter, 1985).

As an alternative to the same-candidates definition, the similar-schools definition of comparability is worth considering for the maintenance of standards between successive examinations on an unchanged syllabus. Indeed, since many schools enter candidates for the same examination in successive years, a *same-schools* definition is possible. With this approach, comparability between two examinations set successively upon the same syllabus would be defined as follows:

Two successive examinations on the same syllabus have comparable standards if two groups of candidates who attend the same schools receive grades which are identically distributed after studying the syllabus and taking the examinations.

The implicit assumptions behind the use of this definition to maintain standards between successive examinations on an unchanging syllabus are that the schools do not change during the period between the two examinations in terms of the school variables affecting pupils' attainments and that the use of the same schools also controls effectively for relevant student variables. It would be possible to research the extent to which these assumptions hold in general. If it was found that they held to an extent which was judged sufficient, then the same-schools definition of comparability would be practically much easier for the examining boards to use than the same-candidates definition. Indeed, the boards already implicitly use the similar-schools definition to maintain standards when they compare the grade distributions of the current examination with those from the previous one. However, the degree of emphasis which is given to this approach varies between the boards and none of them uses it *in place of* awarders' judgements.

One final consideration about the use of exclusively statistical approaches to the maintenance of standards is worth mentioning. This is the transparency of the awarding process. In general, to use statistical methods is to adopt procedures which many people find difficult to understand and of which some people are suspicious. To say that a group of experts (the awarders) judge the quality of candidates' work to award the grades appears comprehensible and reassuring to a lay audience of pupils, parents, teachers and other interested parties. That it, in fact, appeals to a process of judgement which, in detail, is barely understood at all is obscured by the everyday familiarity of the process of making value judgements.

Setting standards

It was argued in the preceding section that statistical alternatives to awarders' value judgements might be viable for maintaining standards between successive examinations on an unchanged syllabus. However, this leaves the problem of setting grade standards for the first examination on a new syllabus or after a revision to a syllabus. Here, the central arguments of this chapter apply and there is no theoretically coherent alternative to the use of value judgements which take account both of *what* is assessed and the *quality* of pupils' performances.

The qualifications needed by those making such judgements are twofold. First, the section on the "nature of value judgements" implies that the awarders must share tacit standards, based upon guild knowledge, with all those making awards in other syllabuses in the same subject and also share more general tacit standards and guild knowledge with those making awards on examinations in other subjects which report results in terms of the same grade scale. Second, the tacit standards adopted by the awarders must reflect the views of the wider group of examination users (see the section on socially defined comparability) sufficiently to be accepted by them as comparable with those applied to other examinations reporting on the same scale of grades.

This leaves open the issue of where the tacit standards come from in the first place. They are perhaps best understood as a dynamic norm established within the teaching profession considered as an identifiable group within society (see Brown, 1988, for a discussion of norms within social groups). The norm is dynamic because it clearly changes over time as the curriculum, in its widest sense, develops. It is rooted in teachers' professional experience (of both their pupils' attainments and the way in which these are rewarded in examinations), discussion with their colleagues and contact with educational thinking and society in general. The dynamic norm by which examination standards are defined does not, therefore, represent an objective yardstick with which changes in the performance of candidates over a long period of time can be measured. It has, however, been the basis of public examination standards which have proved sufficiently stable to be widely accepted as part of selection processes in the educational and vocational worlds for many years.

CHAPTER 6

Group Differences and Bias in Assessment

Harvey Goldstein

Institute of Education, London, England

INTRODUCTION

This chapter is about interpreting differences in educational performance between groups, especially those defined by ethnicity, gender and class.

One of the difficulties in this area is that, in trying to communicate with the non-specialist, assessment experts often use technical terms such as "bias" which have everyday, and less precise, meanings. It is hardly surprising that such an enterprise is fraught with difficulties, even when the attempt is genuinely one of honest communication rather than compliance with custom or even subtle indoctrination. I shall try to avoid such confusions wherever possible by using less ambiguous terminology.

THE NOTION OF BIAS

Central to the concerns of this chapter is the term 'test bias' which has been used by psychometricians and test constructors to refer to group differences which have nothing necessarily to do with the common understanding of bias as distortion. Some practitioners (see for example Shepard *et al.*, 1981) have attempted to inject more precision and acceptability into this term by defining test bias as follows:

A test (or item) is biased if "two individuals *with equal ability* but from different groups do not have the same probability of success" on the test or item (my italics).

Such a definition clouds the issue even further since it falls back upon another term "ability" which is undefined and indeed can only be defined in terms of other

Assessment: Problems, Developments and Statistical Issues.
Edited by H. Goldstein and T. Lewis.

tests (or items) which do not exhibit "bias", with an inevitable circularity. Partly for this reason the term "differential item functioning" (DIF) has come to be adopted for test items which exhibit "discrepant" behaviour. In the context of item response models, as pointed out in Chapter 4, such behaviour can be regarded as an indication of a poorly fitting model. On the other hand, proponents of strong "item response *theory*" would regard this as evidence that the item should be discarded, and this seems to be the sentiment which lies behind the above quotation.

GROUP DIFFERENCES

There is no shortage of empirical evidence for group differences. Large scale surveys (for example Davie *et al.*, 1972) have demonstrated higher performances and rates of progress among middle-class as opposed to working-class children, differences among ethnic groups and between males and females. Most of the empirical work, however, has been carried out in the area of gender. For one thing, the categories are unambiguous and not subject to definitional changes, and for another, in many countries there is clear equal opportunities legislation which has forced a close examination of assessment procedures. I shall not attempt to summarise the empirical evidence here (but see Stobart *et al.*, 1992a, b; Gipps and Murphy, 1994). Rather I want to explore methodological issues and in particular to attempt a typology of approaches to the study and interpretation of group differences.

THE GOLDEN RULE CASE

To illustrate the complexity of the issues I shall describe briefly the history of a dispute which arose between the Golden Rule Insurance Company of Illinois and Educational Testing Service (ETS). The insurance company and ETS, following debate, agreed to adopt a policy of item selection for insurance entry tests which minimised Black–White differences. It worked by test constructors choosing a pool of items, all of which satisfied standard criteria for test inclusion such as face validity and acceptable discrimination parameters. From this pool the final selection was made by choosing those items which produced the smallest (on average) differences between Blacks and Whites. The procedure was discontinued after some years, with ETS claiming that it was inappropriate.

The debate which followed (Goldstein, 1989; Linn and Drasgow, 1987; Anrig,1988) raised a number of technical issues, and the predominant reaction from psychometricians was that technical criteria alone, such as high reliability or correlational validity, should determine test content: one way or another, rather than social or political criteria; these criteria should play no part—even after the technical criteria had been exhausted. Thus the key dispute was between an established psychometric tradition of seeking technical solutions to problems of equity, and the proposition that it is also legitimate to seek a political or social solution. In the case of Golden Rule, the political consideration arrived once technical procedures were exhausted, but there is no reason why political or social desiderata could not be introduced at an earlier stage of the process.

There is always a choice to be made. Criteria for judging whether group differences on an assessment instrument constitute undesirable "bias" must ultimately derive from judgements which are conditioned by existing cultural and political constraints, whether these constraints involve pressures to adhere to historical precedents or to adopt changing mores. In the next section I explore this further in relation to gender differences.

ITEM ANALYSIS AND SELECTION

The standard psychometric approach to the production of a test or assessment is first of all to devise items or questions from which a final selection will be made. The procedure for devising such items will vary, and involves elements of judgement by "experts", and modification of existing items. It is at this stage that subjective elements of choice will enter, usually in an uncontrolled manner. Where items can be piloted prior to a final selection, the test constructor will rely on mainly statistical procedures to eliminate "unsuitable" items and the various psychometric models for doing this are discussed in Chapter 4. The textbooks on test construction, however, typically pay little attention to the problems of initial choice, preferring to devote most attention to the techniques of "item analysis" or "item response model" analysis.

Following an initial selection of an item "pool", in relation to equity considerations items will be screened for obvious biases, looking, for example, at gender or racial stereotypes in language or pictures. Following this the patterns of responses to the items will be examined in detail. This stage has two principal aims. The first is to eliminate items which contain little information, for example those which everyone gets correct or fails. The second aim is to identify "discrepant" items prior to eliminating or modifying them. Chapter 4 discussed this aim in relation to dimensionality and test equating. Here I shall look at it in the light of group differences.

In the absence of any external criterion against which to evaluate the test items (which I shall return to below), only the relationships among the item responses themselves are usable. To judge whether any single item is a candidate for exclusion or modification, the standard assumption is that all the items should in fact be measuring the "same underlying attribute"; unidimensional in the sense described in Chapter 4. It implies that, apart from chance fluctuations, the response on any one item can be fully predicted from the responses on the remainder. For the simple "identity link" model this leads to an examination of the correlation between each item and the total score derived from summing the item responses. Items with high correlations are said to have high "discriminations" and those items with significantly low values are then candidates for further study.

The difficulty with this procedure is that its results are sensitive to the initial choice of items as Chapter 4 pointed out. To recapitulate, suppose that a test of reading comprehension measures two underlying attributes. Suppose also that only a few items in the test actually reflect one of these attributes. The subsequent item analysis will then tend to assign those items low correlations simply on the grounds that they are different from the majority. Excluding them from the final test will help to ensure that test only measures a single attribute. The problem is that this may not be what is required. In more complex cases, where there are several attributes involved, the item analysis procedure cannot be guaranteed to produce anything sensible at all. The testing textbooks, by and large, attempt to make a virtue out of this unfortunate necessity by declaring that all tests have to reflect only a single underlying attribute in order to have legitimacy. What this really means is that the set of procedures used requires the assumption that a test reflects a single underlying attribute in order to have any logical validity. Such a requirement, however, is a very strong restriction on any assessment instrument.

VALIDITY

The examination and interpretation of group differences is closely bound up with notions of validity. Loosely speaking, an assessment's validity is the extent to which it measures what it claims to measure. Sometimes validity is measured in terms of the correlation between a new test and an old established one—the higher the correlation the higher the validity. Sometimes a test is correlated with an external criterion which is supposed to be itself a valid measure. In addition or instead of, such correlational measures, the items are judged by those designing or using them more or less subjectively in terms of their fitness for purpose.

All of these procedures suffer from similar underlying problems. In the case of the external criterion measures the strong assumption has to be made that the criterion itself possess a high validity. In fact, in the case of a new test replacing an old one, it seems difficult to justify the former on the grounds of a high correlation with the latter which presumably is felt to possess important deficiencies. It is not difficult to see how, if this kind of criterion is adopted seriously, historically determined group differences can come to be perpetuated. Thus, for example, given the well-documented (Gould, 1981) evidence for ethnic differences in early tests of ability, the ethnic differences observed in current instruments may be, at least in part, the consequence of applying this psychometric constraint when developing new tests.

In the case of the "face validity" judgements of test constructors and users, there has to be an assumption that such people have a valid understanding of how a test item or question relates to an imperfectly articulated attribute. Yet test constructors and users are themselves conditioned in their expectations by existing evidence. If they believe, for whatever reasons, that boys really do better, for example, on spatial mathematics items, it is hardly surprising if they then tend to reject those spatial items which favour girls. Gould (1981) provides a good example of a similar mechanism operating among the late nineteenth century craniometrists, meticulous scientists who nevertheless were strongly influenced by their cultural expectations when forming judgements. Again, therefore, it is easy to see how historically determined patterns can persist. It would be interesting to have detailed evidence of decisions made by test constructors confronted with having to choose between items in a test, some of which favoured one group and some another. We have little systematic evidence of how decisions are taken in such cases, and the fact of such choices having been made is almost never recorded.

If the above arguments are accepted, they cast some doubt upon the validity of historical comparisons of group differences. Since the assessments used generally change over time, it may well be the case that the new assessments have built in some of the previous observed group differences in order to satisfy "validity" requirements. On the other hand, there may be situations where external pressures for reform of assessment systems have altered group differences. Thus, for example, there is evidence that a change from essay-type questions to multiple-choice questions in English public examinations favoured boys over girls (Murphy, 1982). This also raises more general issues of how changes over time can be interpreted, but I will not explore these here.

BIAS AND GROUP DIFFERENCES

I have alluded to the definition problem of bias and why the standard psychometric criterion is inadequate. The question naturally arises as to whether there is any other sense in which the term can be used.

If an assessment is designed with items that are set in contexts familiar to one group but not to another, we would, in common usage, normally think of such an instrument as biased in the sense of being "unfair". A possible defence against such bias would be that the context was germane or appropriate to that which was being assessed, and hence legitimate: any difference between groups would then convey useful information rather than reflect merely an irrelevant feature of the assessment instrument. In reality we would expect many assessments to reflect "bias" in the above sense as well as relevant group differences. In what follows, therefore, the term "group difference" refers simply to the existence of a difference whereas the term "bias" will be used to refer to a difference which is strictly irrelevant to the attribute being assessed.

An interesting debate has been pursued in mathematics education, where there is an acceptance that problems should be presented in "real life" contexts. For example, if such contexts are more familiar to boys than girls, then the former will tend to do better than the latter for this reason. Yet a change in context, although not in the intention of the measuring instrument, may well affect the relative performance of boys and girls. The question immediately arises as to what contexts to use, and I will return to this issue in the next section.

A related set of issues is raised by an example given by Murphy (1989). Here, a problem on comparing conductivities of different materials was set in the context of using the materials in clothes to be worn when walking on hills. Whereas the boys tended to ignore the "real life" setting, the girls were concerned with what would happen if it rained and the clothes got wet, etc. In this case, the procedure for judging the assessment might be said to be biased against the girls (or any other group with such responses) if it failed to give credit for such observations. Murphy (1991) elaborates on the different types of solutions boys and girls bring to problem-solving tasks, in particular their relative unwillingness to abandon an ostensible "real life" context in favour of an abstracted technical issue. The issue then is to decide just what is relevant to judging a response.

In these examples there is no clear-cut procedure for judging whether bias exists or not, and a similar situation seems to exist with other forms of assessment. There would certainly seem to be a case for using the term when there is a clear intention on the part of the test constructor (which might have unconscious origins) to produce particular group differences. Short of this, however, it is difficult to sustain the use of this term and it would in most cases be preferable to drop its use in favour of referring to group differences or differential performance.

Returning to the psychometric definition of "bias" quoted earlier, that

A test (or item) is biased if, "two individuals *with equal ability* but from different groups do not have the same probability of success' on the test or item (my italics).

we can see not only that it has an inherent circularity, but also that its use can result in a subtle obfuscation of important issues. The use of the term "ability", or indeed any other term such as "attainment", is fraught with problems. Since all ability measures themselves incorporate group differences it is difficult to see how any other assessment can be judged against them. Moreover, different ability tests will incorporate differently sized group differences.

When constructing and examining assessments it is very important to eliminate *irrelevant* features which may result in group differences. A wealth of experience is now available to do this and there are various guidelines, but a decision about relevance is not always easy. Furthermore, what may be irrelevant in terms of gender differences may not be so when considering racial differences and so forth. What is required, in the interests of equity and accountability, is an adequate description for any given assessment of the criteria used in its construction and the criteria used for item or question selection. It would be an important step forward were these to be incorporated into the repertoire of all existing assessment construction procedures.

DESIGNING DIFFERENCES

Where group differences are shown by a particular assessment, these should be viewed as characteristics of the particular assessment itself, or rather of the interaction between the assessment and the groups. In discussing the interpretation of assessments it is important, first, to separate out what might be termed a research activity, where the aim is to investigate why certain group differences exist. In the area of gender differences there has been work in this area (see for example Foxman *et al.*, 1990). On the other hand, where assessments are used for selection or certification, the issue is more pressing. To illustrate the problems, consider the following case of selecting children for secondary education.

The British Equal Opportunities Commission (EOC, 1982) has stated that "any allocation made (to schools or streams) should be *solely on the grounds of ability*" and that separate sex norms should not be used (my italics). Several Local Education Authorities (LEAs) have had to abide by the letter of this guideline (Goldstein, 1986a), but the real issue is more complex.

The situation in several LEAs operating selection for secondary education at 11 years into grammar or secondary modern schools has been that the standard tests, usually of non-verbal or verbal ability, produce higher mean scores for girls than boys. Thus, an LEA which wished to select equal proportions of boys and girls for grammar school education would be obliged to have separate cut-off points on a common test, with that for girls being higher than that for boys. One consequence is that some girls will fail to get into grammar schools even though they have scored higher than some boys who are selected. It is for this reason that the EOC finally ruled in favour of common norms and cut-off points.

The real problem, however, is that an LEA which wished to subvert the EOC's intention could ask a test constructor to design a new test which, as far as possible, equalised the score distributions for girls and boys. The use of such a test would presumably not transgress EOC guidelines so long as a common cut-off was used. Yet it could achieve the same end result as having separate norms. Thus, simple attempts to legislate in this area may appear to address the issue without necessarily resolving it. It would be perfectly possible of course to require all selection procedures to select (on average) equal proportions of boys and girls. This might be justified on social, educational, political or administrative grounds and there may well be problems in making it become generally acceptable.

The point I am making is that test construction technicalities are of secondary importance compared to the choice of desired outcome. Furthermore, when explicit discussion of outcomes is absent there is always a set of *implicit* decisions being made which will determine the outcomes, and these are not necessarily the same as those which might follow from a rational debate about outcomes.

In some educational systems it may be possible to have useful discussions about the acceptability of designing assessments with specifically determined gender differences (or rather lack of them), because there already exists experience of legislation on equal gender opportunities. The issue when applied to ethnic minorities, however, would seem to raise greater problems. The Golden Rule case is perhaps the nearest public debate which has occurred on this, and that clearly raised uncomfortable issues for the testing profession. A further difficulty arises if we wish to design assessments *simultaneously* satisfying several specific group differences, say for gender and ethnic status.

IMPLICATIONS

The thrust of my argument has been that the business of constructing assessment instruments is a complex one involving social and political assumptions as well as technical manipulations. I would also suggest that, because of the difficult issues this raises, it has not been easy to provoke a public debate. Added to this is the power and influence of the "testing industry", especially in the USA and its scientific dependencies (see Chapter 11). This industry thrives, at least partly, on the need to invent and maintain sophisticated procedures for producing and modifying tests and assessments which underpin a large part of education and training. Indeed, the typical response, as in the Golden Rule case, when faced with a "political" challenge, is to attempt to devise ever more sophisticated technical devices to deal with it.

Procedures such as that used in the Golden Rule case could usefully be considered for incorporation into standard assessment construction techniques, including single-occasion examinations. There seems to be no reason why the principle of trying to abolish (or otherwise constrain) group differences should be limited to gender and ethnic groups; individuals can be classified in any number of ways. Of course, practical considerations will be important, as will current political priorities. One merit of having a debate about such proposals is that it would stimulate an appraisal of existing assessments and their characteristics. In addition, as I have already mentioned, recommended procedures for test constructors could usefully incorporate the requirement that a record is kept of the assumptions about group differences which have been used to discard or incorporate items or questions. These assumptions are often deeply buried in cultural norms, including the way in which existing assessments behave, but the attempt to make them explicit should provide long-term benefits.

I am suggesting that traditional ideas of "fairness", "bias" and "equity" in assessment derive from an assumption that, although difficult, there is, in principle, a procedure whereby particular groups are not disadvantaged by the form of any particular instrument. The problem is that there is no external criterion of fairness, and choices have to be made which inevitably include social, cultural and political values. To use terms such as "bias" and "fairness", rather than simply referring to group differences, implies a belief in an objective criterion for judgement. In many ways this is to make the same mistake as those who rely on psychometric notions of objectivity; that is, to assume that some kind of "expert" or "technical" judgement is available and we simply need to work harder in order to find it. This is not to belittle the important attempts that have been made to eliminate from assessments obvious examples of stereotypes and narrow contexts about which there is a broad consensus of opinion. That series of endeavours indicates what it is generally agreed should not be done, but it does not provide criteria for making final choices. Those criteria will be a matter for debate in which differing individual values will play a major role.

There also is another area where useful research could be carried out. Because the outcomes of assessment affect self-image and the views of others such as teachers about attainment, we can envisage setting up studies which deliberately modify assessments in order to enhance the performance of different groups. Thus, for example, mathematics assessments which included items tending to favour girls could be contrasted with those which did not. In an experimental situation, their effects on student progress could be studied.

Such research into group differences should make it possible to achieve a greater understanding of why groups differ, and to investigate the results of deliberately tailoring assessments to achieve particular outcomes. Such knowledge may not tell us precisely what to do, but it ought to make the consequences of any assessment decision more predictable. It may also cause us to change our minds about priorities. After all, perhaps we should accept that Murphy's girls were right to be concerned with the real-life consequences of their designs, and that less priority should be given to formal as opposed to contextual understanding. Such a choice might or might not, in the long run, advantage girls, but at least this and other similar investigations will have forced us to examine what we are assessing from a different cultural, social or political standpoint and to allow ourselves the opportunity to benefit from such experiences.

ACKNOWLEDGEMENTS

I am grateful to Caroline Gipps for her comments on a draft.

CHAPTER 7

Moderation Procedures and the Maintenance of Assessment Standards

Toby Lewis

University of East Anglia, Norwich, England

INTRODUCTION

We introduce this chapter with an example of a public discussion of an assessment exercise in England and Wales.

"Criticism of English Test Marks Floods In

"Hundreds of schools were returning their English national test scripts to the examination boards this week following widespread complaints of 'negative' marking, 'unfairness' and 'farce'.

"English specialists and head-teachers have claimed the marking of the tests for 14-year-olds has been too rigid. Too many bright children have been given low test levels, and some less able students have done better than their more talented peers, they claim. Also, they say the standards applied to this year's tests, marked by external examiners, were not the same as those used last year when the tests were marked by teachers and scrutinised by external examiners. ...

"Complaints continued to flood into the offices of the boards, the exam advisers' quango, the School Curriculum and Assessment Authority [SCAA], and *The TES* [*Times Educational Supplement*]. The National Association for the Teaching of English (NATE) claimed to be receiving a complaint every eight minutes. ...

"At Devonport High School for Girls, a grammar school in Devon, headteacher Barbara Dunball, called the English key stage 3 [KS3] results [at 14 years] 'manifest nonsense'. Last year more than 40 of the 96 Devonport girls who took the KS3 English tests achieved level 7, which meant they had exceeded the target expected of their age group. This year girls of the same ability were awarded only three level 7s. ...

"Many who contacted *The TES* were worried about *consistency of standards* [our italics]. Alistair Darbey, a head of English from Stokesley in North Yorkshire, wrote: 'What do we do about the fact that we are being asked to believe that the same standards

Assessment: Problems, Developments and Statistical Issues.
Edited by H. Goldstein and T. Lewis.

are being applied each year, when clearly they are not? We got it right last year but wrong this year. How are we to feel any confidence in this system at all? . . . I want to say to my students who achieved level 4, on these tests, that they show every sign of being able to reach, at least [at 16 years], a grade C at GCSE [the General Certificate of Secondary Education], if these are the standards by which we are to judge them at KS3. After this, will they believe me?'" (*TES* (*Times Education Supplement*), 23.6.95).

"Whose Mark is on This Failure?

by Alastair West

"Where does responsibility rest for what looks to be an expensive debacle over the key stage 3 [KS 3] English tests? As initial incredulity at the scale of the problem gives way to speculation about its cause, early suspicion points to the exercise having unravelled at the marking stage.

"Rumours abound as to markers' lack of experience and expertise. Returned scripts reveal widespread slavish adherence to the mark scheme and failure to use the exemplar material. There is deep scepticism as to the quality and extent of supervision received by markers. . . .

"Its front-line position will ensure that SCAA ... will collect brickbats. This is understandable without being welcome. That body has worked hard and successfully to establish a more positive professional climate. It may influence, but it does not make policy. No, the 'onlie begetter' of this KS3 tragi-comedy is to be found elsewhere.

"Last summer, the Secretary of State [for Education] removed the requirement for any statutory audit of teacher assessment. Schools were exhorted to participate in voluntary arrangements, but no funding was available to support them. Enormous sums, however, were found for the external marking of tests.

"This year's teacher assessment results in KS 3 English are likely to be more secure than the test results, given the quality of test marking. But *without moderation there can be no proof* [our italics]. There is a grim irony now to Dearing's recommendation that test and teacher assessment results have equal standing. Government policy has ensured the unreliability of both. If there are to be tests, they should be marked by classroom teachers and *supported by rigorous moderation procedures* [our italics] for both the tests and teacher assessment. . . .

Dr Alastair West is chair of the National Association for the Teaching of English." (*TES*, 30.6.95).

In the grading context, where summative assessment is used to provide certification for the information of the outside world, assurance is required that the assessment is objective and "fair". There are two issues here, the content of an examination (or other form of assessment) and the marking of the candidates' work.

First, is the examination or test, as set, well designed for its purpose? Is it of the "right standard"? Is its standard comparable with those of examinations set for the same purpose by other institutions, or by the same institution at other times? A variety of monitoring procedures are in use in different quarters, aimed at providing assurance on these points. How well do they perform in maintaining standards? What in fact is implied by "maintenance of standards" and by "comparability"? A discussion of these problems is given in Chapter 5 and we shall make some additional comments later in this chapter.

Secondly, however well designed and well set an examination may be, it will fail in its purpose if the marking is defective. Effective moderation is essential for the marking process as well as the setting. This is cogently illustrated by the extracts from the *Times Educational Supplement* of 23 and 30 June 1995 at the start of this chapter.

Our main purpose here is to review some current procedures, with particular reference to the United Kingdom. It is reasonable to look at United Kingdom practice, because the relatively small size of the country and the short distances make it easy to use moderators and external examiners (for universities) from institutions in centres other than the one served.

There are different contexts to be considered, including educational assessments at tertiary level (university and college examinations), at secondary level (school examinations), and professional and vocational qualifications (engineering, medical, architectural, etc.).

SOME CURRENT PROCEDURES

Universities

In the UK, the Committee of Vice-Chancellors and Principals (CVCP) issued in 1986 a Code of Practice for the External Examiner System for First Degree and Taught Master's Courses (CVCP, 1986). This begins with the following statement of "purposes and functions":

1. The purposes of the external examiner system are to ensure, first and most important, that degrees awarded in similar subjects are comparable in standard in all universities in the United Kingdom, though their content does of course vary; and secondly that the assessment system is fair and is fairly operated in the classification of students.
2. In order to achieve these purposes external examiners need to be able:
   - to participate in assessment processes for the award of degrees;
   - to arbitrate or adjudicate on problem cases;
   - to comment and give advice on course content, balance and structure, on degree schemes, and on assessment processes.

The code continues with sections on "formal requirements", "selection and appointment", "period of service", "participation in assessment procedures", and "written reports".

In fact, the usual practice at any particular university for maintaining standards in the examinations in a particular subject and for ensuring "fairness" in the operation of the assessment system is to appoint an external examiner in that subject from another university, generally for a term of three or sometimes four years. He or she (for brevity we will say "he") first vets the examination papers in draft and, if he judges necessary, revises them with regard to such aspects as the following:

- comparability with standards of corresponding examinations at other universities;
- ease or difficulty of individual questions;
- coverage of syllabus;
- time required for candidates to answer the questions in relation to the duration of the examination;
- equal weight of questions;
- details of wording and presentation, to secure clarity, avoidance of ambiguity, etc.

After the examination has been held he reviews the marking, possibly of all the scripts, depending on the number of candidates, certainly of a substantial sample. This sample will include the scripts of all candidates whose performance is provisionally on or near a critical grading borderline. He may decide to change the marks of an individual candidate if he judges that too much or too little credit has been given by the marker to some aspect of the script. He may also rescale the entire set of marks for all the candidates, if he considers that this is required in the interests of consistency of standards.

Following this there is the examiners' meeting, where the external examiner (or external examiners if, to cover several component subjects, there is more than one) acts as final arbiter in the decisions on award of degree classes, etc. (This is a slightly grey area; at one university, for example, the stated duties of the external examiner include "being influential in cases of disagreement over marking and classification, where the external examiner's views carry particular weight"; at another university it is similarly stated that "any disagreement . . . shall be resolved by discussion by the Board of Examiners in which the views of the external examiner shall carry particular weight". Yet there are other universities where it is laid down that, in the case of disagreement between the members of the board of examiners, the external examiner's recommendation shall be adopted.) The external examiner has the power to recommend changes to provisional gradings given by the internal examiners, if the quality of performance of a borderline candidate appears to him not fully reflected (one way or the other) in the candidate's marks. Before the close of the examiners' meeting he will give his comments on the results, the examinations, the marking, the syllabuses, etc. and possibly offer advice on points to be taken on board for the next round of the examination process. At some universities he may give a *viva voce* examination to a borderline candidate if this is needed to reach a clear decision.

As an essential part of his duties, he is required to submit written reports, both annually and at the end of his period of office. He may also be called upon to play a role if the integrity of the grading operation is put in question through actual or suspected malpractice by one of the parties involved: a candidate, an examiner, even an institution; this is infrequent, but it can occur.

We have referred above to the CVCP code of practice. Individual universities issue their own lists of duties of external examiners, varying in length and explicitness. These reflect, in greater or lesser detail, the actualities of external examining described above. Universities also specify what coverage they require in the external examiner's annual and final reports; these specifications likewise vary considerably in detail and amplitude from one university to another.

Professional institutions and their qualifications

While some professional institutions use university examinations to meet their academic requirements (as distinct from, say, evidence of work experience), many institutions use examinations either set by themselves or specially set for them —see the account below of the Engineering Council and its work. In some cases the institution's examinations and the appropriate university examinations are both available as alternative routes towards qualification.

There is a variety of arrangements in place for moderating these professional examinations. Some are of impressive thoroughness. Consider, for example, the system set up by the Engineering Council for maintaining the standards of their examinations and ensuring fairness in the assessment of candidates.

The Engineering Council (EC) was established by Royal Charter (1981) "to advance education in, and to promote the science and practice of, engineering (including relevant technology) . . . in the United Kingdom". Candidates for chartered status in particular institutions, such as the Institution of Electrical Engineers or the Institution of Chemical Engineers, can satisfy the academic requirements either by university degree or by the EC's examinations, which are of British honours degree standard. Upwards of 40 subjects are covered by these examinations, which are grouped into *divisions* of five or six "like" or related subjects.

For each subject a *moderator*, typically a senior figure from a university engineering department, is nominated by the relevant institution and approved by a committee of the EC. A *chief examiner* for the subject is appointed in the same way. The chief examiner is responsible for setting the relevant paper, to the standard prescribed by the syllabus. This paper is vetted, possibly revised and finally approved by the moderator.

In the next stage, the moderators and chief examiners for the subjects in each division meet to consider and possibly revise the five or six relevant papers. Next, the chief examiner (say "he") appoints one of the team of examiners working under him who, from the proof of the paper, writes solutions which are checked against the chief examiner's solutions.

At the marking stage, the chief examiner allots a proportion of the scripts to each of his examiners, and checks their marking. The moderator then receives a large sample of the marked scripts from the chief examiner and ensures that the marking is fair, possibly revising marks in the process. In the case of any candidate whose overall examination result depends critically on a borderline mark in one subject, the moderator is required to review that script and decide whether a pass is justified. Finally, the moderators and chief examiners in each division meet again and review the results in that division. Their recommendations are passed to the *board of moderators* for ultimate consideration.

An example of a similarly thorough moderation process in a professional area different from engineering is provided by the Chartered Institute of Management Accountants (CIMA). The examination setting part of the process is described in detail in the paper, *Setting CIMA's Examinations* by Jean Elley, Deputy Director, Examinations Technical, CIMA, reproduced here (in slightly abridged form) as Appendix A, by kind permission of CIMA (CIMA, 1995).

Turning to another professional area, architecture, the Royal Institute of British Architects (RIBA) operates a two-track examination system. A candidate for RIBA membership may either take the RIBA examination in architecture (Parts 1 and 2, involving the equivalent of three years and two years full-time study) or, more commonly, may take equivalent examinations as students in a recognised school of architecture, and gain formal exemption from Part 1 and/or Part 2 of the RIBA examination. There are nearly 40 architecture schools in the United Kingdom with courses recognised for exemption. Recognition is based on a rigorous assessment process, and must be renewed every five years: it is carried out by a joint educational validating panel of the RIBA and the Architects Registration Council of the UK (ARCUK). While each school has its own internal examining arrangements, all the schools appoint external examiners, whose brief is to ensure that the school's internal assessment meets the requirements of the RIBA. Currently, RIBA is considering possible steps to further ensure the maintenance of standards; these include the formulation of guidelines for the appointment of external examiners by schools, the promotion of cross-contact between examiners in different schools, and the encouraging of schools to rotate their examiners. A research project on the role and functioning of external examiners in the different schools is currently being conducted.

A major UK system of work-related qualifications: City and Guilds

The City and Guilds of London Institute, founded in 1878, offers an extensive range of technical and vocational qualifications at a progression of levels. Certification is awarded in over 500 subjects from a wide field of some 30 different main areas; they include, for example, such diverse areas as art and design, agriculture and horticulture, catering, electrical and electronic engineering, furniture production, tourism, printing, textile manufacture, vehicle maintenance, office technology and information processing (C & G, 1995). Much of the assessment is carried out by *centres*, which are those educational training establishments which have been approved by C & G to offer assessments leading to C & G qualifications; at the same time, there are instances where elements of the assessment are undertaken directly by C & G. Typically a centre is a college, a school, or a private training provider; it could be a large company with its own training department. Several different methods of assessment are used, reflecting the wide variety of certification contexts. These methods include, among others, observation of performance in the workplace; observation of performance under controlled test conditions; appraisal of products (e.g. artefacts and designs); and written tests which may be open response or multiple choice (C & G, 1994).

As regards maintenance of standards and fair assessment of candidates in this complex system, City and Guilds operate detailed mechanisms for what they call "quality assurance", aimed in their words at ensuring that "the consistency, integrity and quality of City and Guilds assessment and certification procedures and practices are maintained". Where assessment is wholly carried out by the centre,

"The key roles [in quality assurance] are the:

(a) assessor	the person appointed by the centre to assess candidate achievement
(b) internal verifier	the person at a centre who co-ordinates the assessment process
(c) external verifier	the person appointed by City and Guilds to verify the assessment process
(d) C & G officers	the staff who administer the scheme" (C & G, 1994).

Briefly, a centre intending to offer C & G awards must satisfy an approval process; and when it is in operation following approval, the quality of its assessment practice requires verification (termed *monitoring* by C & G). Centre approval and ongoing monitoring are carried out by trained external verifiers appointed by C & G. Internal verifiers are appointed by a centre if it has two or more assessors for an award. Checking assessors' judgements of candidates' performance (termed *moderating* by C & G) is carried out by the external verifier, together with any internal verifier who may be appointed.

Public examinations at secondary level

We consider here the General Certificate of Education (GCE) A-level examinations, which are the main means in UK of selecting students for higher education. The moderating procedures operated by the several examining boards which offer these examinations are very thorough.

In the particular case of one examination board, the University of London Examinations and Assessment Council (ULEAC), the responsibility for a particular paper in the modular syllabus rests with a principal examiner, who is assisted by a number of assistant examiners grouped, if the numbers are large, under team leaders. Over a number of principal examiners is a chief examiner, responsible for a complete syllabus; and over a number of chief examiners is a chair of examiners responsible for maintaining standards in a subject or group of subjects across different syllabuses, and reporting to the chief executive. The revision meetings, held to ensure that the draft question papers and mark schemes meet the requirements of the syllabus and scheme of assessment, include as participants one teacher reviser and one university reviser. At the marking stage, samples of scripts are photocopied and check-marked by all examiners to ensure that they are each interpreting the mark scheme consistently. Separate staff at headquarters go through every script to check that marks have been totalled correctly, every page of work has been marked, etc. In a final check, team leaders take random samples of scripts and check that each individual examiner has carried out their marking satisfactorily; if necessary, remarking of scripts is undertaken.

We should mention two other procedures which have been used by GCE examining boards for purposes of maintaining standards. One is where a set of past scripts from, say, 15 years ago are re-marked in red light (so that the original marking cannot be seen) by examiners of the current generation. This gives information on possible changes of marking standards over time. It has generally been found that the present assessments do not differ significantly from the past ones. The second procedure is a case of co-ordination between examining boards. A particular board's papers and marks for a particular subject in a particular year are looked at by examiners in that subject from the other boards; each board takes its turn. This exercise has proved very valuable, especially when carried out for a new syllabus.

THE RATIONALE FOR EXTERNAL EXAMINING IN EDUCATIONAL ASSESSMENT

The CVCP Code of Practice, quoted above, states that "the first and most important" purpose of the external examiner system is to ensure "that degrees awarded in similar subjects are comparable in standard in all universities in the UK". The assumptions which underlie the external examiner system, in the context of United Kingdom universities, have been critically analysed by Warren Piper (1994). He discusses the following key issues among others:

- The desirability or otherwise of comparable standards.
- The extent to which universities constitute a single system for the award of degrees.
- The reliability of examination marking and the variation between marks awarded by different examiners of the same work.
- The reliability of an external examiner's own judgements.
- The variability of grade or degree profiles between different universities and between different subjects or different faculties or schools within a university.
- The problem posed by joint and modular degree schemes.

Warren Piper also gives a useful list of some 120 references. It is clear that the maintenance of comparable standards is not as achievable nor indeed as well defined as is often conventionally assumed.

The issues of setting and maintaining standards and of defining comparability of standards are analysed in depth by Cresswell in Chapter 5, with particular reference to public examinations at secondary level. Similar inferences can be drawn regarding the limitations on what an external examiner system can achieve in practice.

We do not deny these limitations, but there is still a need for an external checking process. We refer back to the report on the 1995 English Key Stage 3 test results quoted in the introduction to this chapter. This has a clear moral: it is essential to have a moderation procedure to guard against gross malfunctioning of an assessment system. In our view, this is its overriding purpose. Procedures for trying to ensure comparability of standards can be thought of as devices for avoiding extreme events which people would agree represented marked deviations from a common standard. Their function for making precise judgements is open to question.

OTHER VIEWS ON EXTERNAL EXAMINING

Among other publications on external examining in recent years are the following substantial reports.

- *Handbook for External Examiners in Higher Education* (by J. Partington, G. Brown and G. Gordon: CVCP, 1993) is a comprehensive manual for guiding inexperienced external examiners through all aspects of current practice.
- *External Examiners: Changing Roles?* (by H. Silver: CNAA, 1993) is a study of examination boards and external examiners. It is based on observations of examination meetings and discussions with staff and examiners and suggests future directions.
- *The External Examiner System: Possible Futures* (by H. Silver, A. Stennett and R. Williams: QSC, 1995) is a report of a study commissioned by the UK Higher Education Quality Council to examine the effectiveness of the external examiner system. It refers to the *acknowledged central purposes* of the examining system, which we have argued is open to varying interpretations. The report makes a number of recommendations, of which the first two are particularly to be noted:

1. The external examiner system should be retained and strengthened in whatever ways are necessary for it to operate effectively in the future. Other possible changes in higher education, such as any changes in the honours classification system, would only reinforce the need.
2. In the presence of major changes, including scale, diversity, modularisation and semesterisation, the purposes and roles of external examining, which may differ operationally across institutions, should be clarified. The traditional roles of external examining, to ensure comparability of standards and fairness to students need to be reviewed in the light of changes in higher education.

CONCLUSIONS

We have looked at external examining and quality control procedures in several different certification contexts with differing requirements. Maintenance of standards is probably easiest to achieve for a professional qualification, for example in chemical engineering, architecture, etc. where the professional institution may only need to concern itself with its own standards. For a university degree examination in a particular subject, on the other hand, there are issues of comparability with other universities offering the same qualification. Similar considerations apply to public examinations at secondary level such as GCE A level. A different situation arises with a system such as that operated by City and Guilds, offering a great and diverse range of work-related qualifications assessed in a variety of different ways. The extensive arrangements in place for monitoring and moderation are determined in the first place by the "within system" requirements of content validity and reliability, taking account also of such factors as the desirability of divorcing assessment from the mode of learning —a factor hardly present in, say, a university context.

Thus different procedures for the maintenance of standards are appropriate for different certification contexts. The idea of a set of common procedures across all kinds of qualifications is not valid.

ACKNOWLEDGEMENTS

I am grateful to many people for help and information, in particular Mr W. Coach (ICE), Mr G. Cumming (ULEAC), Mrs P. Edwards (RIBA), Mrs J. Gardner (CIMA), Mr B. E. Millicent (EC), Mrs E. Byram and Mr A. Sich (C & G) and Professor T. M. F. Smith (University of Southampton).

APPENDIX A Setting Examinations for the Chartered Institute of Management Accountants (CIMA)

Jean Elley

Deputy Director, Examinations Technical

A SNAPSHOT VIEW

There is a never-ending cycle of activity to bring CIMA's examinations to you. Let me tell you about July for example: at the beginning of the month, the examining teams were just finishing the marking of all your scripts from the May examination; there were four meetings of the Examining Panels to agree the examination question papers for the November examination; and in the middle of the month, the 16 Examiners started to draft the question papers which will appear on candidates' desks next May.

But rather than have that "snapshot" image (rather like a balance sheet, frozen in time), perhaps you would like the fuller story?

A CLOSER VIEW

The Examination Committee has selected and appointed 16 people to act as CIMA's Examiners—one for each subject. It has similarly selected and appointed 16 people to carry out the work of Examination Assessor, again one for each subject. Five Moderators have also been appointed—one for each of the four Stages of the examination, and an overall Chief Moderator. Each of these people, and the CIMA staff involved, has a defined job to do, and there is a strict timetable to which the work is carried out.

DRAFTING THE NOVEMBER EXAMINATION PAPERS

To illustrate this, let us consider the November 1995 session of examinations. Work started on drafting the examination question papers in January. The Examiner for each subject will have considered the framework for his or her question paper—the syllabus itself, how many questions in each section of the paper, the type of question, how they sit together, how they will provide coverage of the syllabus. In doing this, he or she will have considered the shape of the pilot paper and of the May question paper, what advice he/she has already given to students and lecturers via the various means at an Examiner's disposal: syllabus guidance notes, list of recommended reading and previous Examiners' Reports. The Examiner produced a draft question paper, and also some spare questions which could be used as substitutes if necessary, or which could provide the basis for the May 1996 paper. The Examiner also produced answers to the questions—these answers will become the marking scheme. Each draft paper (but only the paper—not the answers) was then given to the subject's Examination Assessor so that the paper could be "tested".

ASSESSING THE DRAFT NOVEMBER PAPERS

The Assessor sits down and answers the questions under simulated examination conditions: he or she is assessing whether the questions are technically correct, whether there might be any errors of fact or omissions from the question (this can come to light if an assumption has to be made rather than a fact or figure deduced), whether the question being asked is at the appropriate level for the Stage and for the likely level of knowledge expected of candidates. Do the questions reflect the aim and weightings of the syllabus? Are they written as clearly as possible? Is it possible to answer the questions properly in the time available—and the mark allocation is as much of a guide to the Assessor here as it is to the candidate later. The answers which the Assessor produces eventually become the published *Suggested Answers.*

The next step in the proceedings was that the Examiner and Examination

Assessor exchange their sets of answers, identify any points where they diverge, talk about the coverage and the way the questions are worded. What appears on the desk in front of a candidate should not in itself be a test: the examination is on the use the candidates make of the information in front of them and the question itself should not be an impenetrable maze through which they have to find their way. The outcome of these discussions was that the draft paper and answers (both sets) were agreed, and the paperwork arrived at the Institute—the date set for this was in early March.

MODERATION PROCESSES

At this point, the paperwork for all four subjects in the Stage was sent to the appropriate Stage Moderator. His, or her, work involves going through each set of paperwork to see that each of the four question papers relates realistically to the practical work of a management accountant; to ensure that the four papers across a Stage will be of an acceptable and common standard and that each gives adequate coverage of its syllabus. That can be said in relatively few words, but the four people concerned devote a considerable amount of time and energy to the detail of this work. Four weeks are allowed, and a "report" on each subject and on the Stage as a whole was provided to the Institute by the middle of April.

Simultaneous with the Stage Moderators' work was the task of typesetting the question papers: Institute staff look after this, using a desk-top publishing operation, so that from here on all those involved see the draft question papers in the format in which they will appear on candidates' desks.

The Stage Moderators' subject reports were given to individual Examiners and Examination Assessors so that they could respond to the points raised: this can sometimes be a small matter of changing a word or two—or a much greater matter of scrapping one or more of the questions and writing new ones to replace them. The Examiner does not necessarily have to give way on all points to the Moderator, but would have to provide a convincing and logical argument in defence of points which a Moderator has queried if he or she wishes to retain them in questions.

While this is happening, the paperwork for all 16 subjects was supplied to the Chief Moderator. His work is in some respects similar to that of the Stage Moderators, but he has to take an overall view and ensure that there is a correct relationship between the Stages, especially for subjects which are related. He reported back to staff within the month prescribed for this work and his reports were supplied to individual Examiners and Assessors. This stage of proceedings also resulted in various refinements being made to the question papers, making guides and Assessors' answers—in a similar way to the reaction to Stage Moderation, the Examiners would have had to argue convincingly if they felt strongly that the Chief Moderator was misguided.

THE EXAMINING PANELS

So, Examiners and Assessors responded to the points made and reached agreement with the Moderators. At the beginning of July there was a meeting of each of the Examining Panels: there are four of these, one for each Stage. They comprise the four Examiners, the four Examination Assessors, the Stage Moderator and the Chief Moderator; each meeting is chaired by a member of the Examination Committee and there are four Institute staff also involved. The main business of each of these day-long meetings was to agree, in a formal and detailed way, the content of each of the four question papers in the Stage. The scene is set by considering information available from a variety of sources—results of previous examinations, any changes to the syllabus or matters relating to the content of the syllabus such as new legislation or newer techniques. Each question paper is looked at in fine detail—content, wording, coverage are all reconsidered by the group before agreement is reached. The outline marking guides are also agreed and the suitability of the Assessors' answers for publication as *Suggested Answers* is also considered. The overarching aim is to ensure that the examination is a fair and meaningful test of the knowledge requirement of management accountants, and the practical application of that knowledge to situations which they might well meet in their working life.

THE OUTCOME

Following each meeting, staff made the final amendments to the proofs of the question papers and a series of detailed checks of the paper started to take place, so that printing could be arranged in good time. Examiners started to make the final changes to the marking schemes so that they would be ready for use in November. And Examination Assessors finalised their answers into what CIMA will publish as *Suggested Answers.*

CHAPTER 8

Errors in Grading and Forensic Issues in Higher Education

David B. McLay

Queen's University, Kingston, Ontario, Canada

THE IMPORTANCE OF ACADEMIC GRADES

Students in post-secondary institutions throughout the past century have no doubt been concerned about the grades on their academic transcripts. In today's climate of economic insecurity, technological change and widespread unemployment, there seems to be more anxiety than ever concerning grades on official transcripts.

The Chronicle of Higher Education Almanac (1994a) reports on a survey of 35 748 faculty members at 392 colleges and universities of North America in the fall and winter of 1989–90. According to this survey, 25.4% of the faculty in publicly funded universities and 48.9% of faculty members in private universities think that "... there is keen competition among most students for high grades". In these circumstances, it is essential that testing of students be conducted fairly and that grading be done equitably. It is also essential to ensure that the work of students is free from cheating and plagiarism. On the one hand, faculty members must employ good forensic techniques, both to discourage academic dishonesty and to detect it when it occurs. There must be well-established protocols for dealing with incidents of academic dishonesty when the evidence is good enough to justify penalties. On the other hand, there must be adequate opportunities for students to appeal against what they perceived to be either unfair examining and grading procedures or unjust penalties. These matters will be the subject of this chapter.

The writer has taught for 40 years in Canadian universities from coast to coast with several years in British Columbia, six years in New Brunswick and over 30 years in Ontario. His own field is experimental physics but for nine years he served as dean of undergraduate studies in a Faculty of Arts and Science with 7000 undergraduate students and over 400 faculty members. In addition, he was involved for 10 years in the Ontario Universities' Council on Admissions and has

Assessment: Problems, Developments and Statistical Issues.
Edited by H. Goldstein and T. Lewis.

served terms on the university senate of universities in both New Brunswick and Ontario. He also has some experience of universities and technical colleges in France, India and The Netherlands. While the experience of North American universities forms the basis for this chapter, it is intended that it will be relevant to institutions of higher learning beyond the shores of North America. For a discussion of related problems in developing countries see Chapter 12.

DIFFICULTIES IN GRADING

At the best of times and with the best will in the world, instructors find it very difficult to set and mark challenging but reasonable assignments and examinations. The difficulties and subtleties involved are thoroughly described in Chapter 10 of Paul Dressel's *Handbook of Academic Evaluation* (Dressel, 1976). His analysis involves essays, laboratory work, multiple-choice instruments, problem-solving, tests of memory, etc. (see also Chapter 4). Conscientious instructors will find themselves agreeing with much of his analysis. As Quellmalz (1990) has pointed out, there are difficulties with setting good essay-type examinations as well as with multiple-choice tests and: "The complexity of designing valid essay examination programs dramatically illustrates the more general problem educators face in integrating testing and instruction."

There is often a problem with consistency in grading, especially in the case of small classes. An older colleague, now deceased, used to talk about his small class of eight female students in their senior year of mathematics at Hunter College in New York. This professor, who related very impersonally with his students, was perturbed to find that his grading yielded seven A grades and one good B grade at the end of the year, a very unusual distribution. After he had submitted his grades with some trepidation, he discovered that his grades were similar to those submitted by other instructors; it was an exceptionally good class! The writer had a similar experience in 1962 at the University of New Brunswick with a class of seven students in honours physics, all of whom were above average and some of whom seemed to be exceptionally good. Their instructors worried about the rather high marks that these students were receiving with respect to previous classes. Their fears were groundless because three of those students ranked in the top 10 of a national prize examination in physics conducted each year in Canada.

A caveat must be added here about statistics, even those for large samples of students. Colleagues of the writer involved in the teaching of classes of the order of 500 students say that there is great variability between different years and that it would be inappropriate to mark assuming a symmetric unimodal distribution such as the normal. One common phenomenon is a "double-hump" in the distribution of grades, possibly due to a constraint in the testing procedures which keeps a fraction of the students below the barrier. A complication due to examination anxiety has been identified by Naylor (1990): "There is sufficient evidence concerning test-taking anxiety to show overwhelmingly that it interferes with optimal performance. Even where highly able students appear to have their performance facilitated by anxiety, it is not inconsistent to state that their performance would be impaired where the difficulty level of the test was beyond their capacity." In this connection see Chapter 1.

On occasions, a superior class which delights the instructors is followed by a class which is deemed by the same instructors to be unresponsive and slow to comprehend the canonical subject matter. Instructors who have been teaching for decades report that there is a surprising variability from one year to another even in classes which have come from a fairly uniform distribution of feeder schools. Also, it appears that many university instructors throughout the world have a negative view of the preparation of their students in written and oral communication skills and in mathematics and quantitative reasoning skills. A Carnegie Foundation international survey of attitudes and characteristics of faculty members in 14 countries (*The Chronicle of Higher Education Almanac*, 1994b) shows that it is only a small minority of the order of 20% who think that their undergraduate students are adequately prepared for university studies.

Another problem with grading standards, not easily resolved, is that faculty members with different academic backgrounds often have very different scales of marking. For example, a professor with a UK Cambridge or Oxford background who is teaching in a North American university may regard a B+ grade as very good indeed and may award this, or possibly an A− at most, to the better students in the class without realising that such marks in North America may well sabotage the eligibility for scholarships of the students concerned. Although the professor in question might have been cautioned about the consequences of mediocre marks on the transcripts of scholarship students, there is often a mindset of propriety in instructors which associates higher marks with undesirable grade inflation.

In normal circumstances in educational institutions, instructors present well-defined courses of enrichment, instruction and stimulation to their students and then test their assimilation and mastery of the content by means of assignments, essays and examinations which are reasonably challenging and relevant to the course. Unfortunately, aberrations are all too frequent both on the part of instructors and through the dishonesty or negligence of students. Sometimes, the situation is made worse by an insensitive and unresponsive institutional administration which does not wish to get involved in contentious and potentially litigious situations. Such dishonest and evil influences as bribery, intimidation and partiality, although rare, cannot be ignored. Bribery may be much more common in some cultures than in others; see Chapter 12.

In this chapter, no attempt will be made to deal with such matters as bribery of instructors, sexual harassment of students by instructors, rewards for sexual favours granted to instructors, etc., even though these do occur occasionally, as well-publicized cases in the media attest. Most educational institutions have codes of conduct which strictly proscribe such dishonest practices and prescribe stiff sanctions including dismissal of instructors and criminal charges. Mystery story writers such as Amanda Cross (1976) in the United States and Colin Dexter (1977) in England, both of whom have extensive experience in universities, find fertile ground in their backgrounds in universities for sensational accounts of corrupt practices of faculty and staff members.

EQUITY IN GRADING OF STUDENTS' WORK

A major problem facing students in obtaining equity in grading is arbitrariness and unfairness on the part of a minority of their instructors. One would think, quite wrongly in many cases, that experienced teachers are so aware of published guidelines and regulations that violations of accepted standards would be very rare. It is astounding to discover blatant violations of instructions and regulations circulated to instructors, especially just before and during examination periods. A common regulation which is frequently disregarded is that instructors inform their students at the beginning of term on what bases the assessments will be made at the end of term. Another often-ignored regulation is that arbitrary changes cannot be made in methods of assessment in midstream unless all the students are in agreement. Another common malpractice of instructors is to circumvent examination schedules by holding examinations and tests at times and in places which are convenient for the instructors but very inconvenient for one or more students. It is surprising how many times the research schedules or even vacation plans of instructors determine the location and timing of examinations and tests. One form of subverting regulations, in which students frequently connive, is the holding of examinations and major tests in a proscribed period just prior to official examination periods. While this practice does not involve any disagreement between instructor and students, it does undermine the integrity of the official periods for review and study and frequently leads to disruptions of the schedules of the other classes taken simultaneously by the students.

Another contentious issue is the unwillingness of instructors to reveal to students the bases of assessments of assignments, essays, examinations and tests or even to return to students the graded submissions. Some instructors appear to have a morbid fear of challenges to their academic judgement and refuse to discuss grading with their students. In some cases the submitted work is returned with no mark at all or with a mark that is not explained by critical comments or annotations. Students are advised in academic regulations to retain all marked work in case of appeals of grades, but sometimes there is nothing on paper to be appealed against except the final mark. Generally, appeal boards will consider charges of unfair evaluations if the graded work is submitted as evidence. One common form of redress is to ask the head of the department concerned to appoint a competent and disinterested examiner to review the assessment independently of the original examiner. This procedure often applies also to final examinations if the student is willing to pay a prescribed fee for a reassessment. In the case that the student's appeal is upheld, all or part of the fee is returned to the student, but if the appeal is denied the fee is forfeited.

In some jurisdictions, the student can obtain a photocopy of the marked examination by paying the prescribed fee. This is usually a last resort after the student has not been able to get a reasonable explanation for a grade from the instructor in question. Although there has been apprehension in educational circles about litigation in the courts concerning academic standards, the courts have so far been unwilling to enter the area of academic judgements. On the whole, educational institutions have created regulations to ensure that students have reasonable recourse to appeal procedures, and more often than not it is a handful of obstinate and unsympathetic instructors who balk at granting rights of appeal to their students.

Instructors may flout institutional regulations with respect to examinations or tests, for example to suit their own personal convenience. It is not unknown for instructors to require their students to show up in the evening for term tests, etc. even when this is difficult. Unfortunately, students complain infrequently. To the credit of the authorities, they usually take the trouble to ensure that no student is disadvantaged under such conditions and, if necessary, annul the test and order a new one under proper conditions. Another stratagem of inconsiderate instructors is to bypass the regular examinations by holding private examinations in a location of their choice. Because good sites have been pre-empted for the official examinations, these unofficial and unsanctioned examinations often take place in rooms that are badly equipped and too cramped; students may have to sit close to one another, with the risk that casual glances round the room can be interpreted as academic dishonesty by looking at a neighbour's work. In one such case, the instructor who was invigilating charged two students with cheating and started proceedings for their expulsion. One of the students in question was able to convince the appeals board that the answer in dispute was derived from material taken in a parallel course not taken by the other student. The instructor, shaken by this evidence, withdrew the charges completely.

A related matter is that of "confidential examinations". These are examinations which are held with tight security measures which involve the collection of all test papers in addition to the students' answer booklets. The aim here is to use the examination again and again without having to invent new test questions (see Chapter 1). There are two unfair aspects of this procedure, one involving appeals about grades and the other involving advice to some but not all of succeeding classes. Students wishing to appeal against what they consider to be an unfair grade will not have access to the questions on which they were tested. In rare cases, the questions are not even made accessible to colleagues asked to consider the appeal. In subsequent years some students may be privy to the nature of the examination questions because they have been coached by a predecessor with either a good memory or with a set of notes written down immediately following the examination. In extreme cases, a "rogue" copy of the examination is known to some students before the examination is written. It happens in some departments that examinations are of the "multiple-choice" variety and these departments may wish to guard carefully their "test banks" of good test questions. The same issue concerns college entrance boards, educational testing agencies (see Chapter 11) and other external bodies who wish to reuse items; in the United States the introduction of "truth in testing" legislation, where examination questions must be made public, has prevented some of the more obvious equity violations.

MINIMISING AND PREVENTING CASES OF ACADEMIC DISHONESTY

Instructors and institutions should adopt procedures that minimise the incentives and/or opportunities for academic dishonesty on the part of students. This is relatively straightforward in the case of official examinations written under good supervision in official examination centres.

The advantage to having all examinations conducted under centralised auspices of the institution or examination board, is that there is a level playing field for all students (and for instructors too, for that matter). It is to the advantage of students, departments and the institution itself to have all examinations conducted in this way.

A new issue in examinations is the computing power of hand-held calculators some of which are very sophisticated. Many courses in economics, engineering, science and statistics require computations during examinations. Programmable calculators are often capable of storing vast amounts of data, including textbook material and sample calculations. One way of dealing with this problem is the prescription of the types of calculators that are permitted. Another way is that the institution concerned provides standard calculators or sets of approved calculators in the examination halls. A further solution is to allow students to bring in textbooks and/or full sets of class notes to minimise the advantages of sophisticated calculators. There are obvious disadvantages to each of the above solutions and this is an active area for further work. Another related problem is the possibility of hidden communication devices worn unobtrusively under clothing. These would allow a good student to coach a weak student, whether for friendship's sake or for remuneration. The rapid development of communication technology raises the spectre of many forms of sophisticated and subtle if not undetectable cheating. In order to minimise academic dishonesty, students need to be warned of the serious nature of offences and the severity of sanctions that can be applied.

Another issue is that of coursework assignments in the form of projects or essays. Collusion and copying are very difficult to detect. It may be possible to detect this by examining work for identical responses. Many courses require the submission of essays which constitute a major part of a course's work. Unfortunately, there are now many "essay banks" which have catalogues of essays that can be purchased. In some cases, their teams of writers are able to write essays on demand which fulfil all the requirements of the assigned topic. The instructor or teacher who reads such essays may find the high quality of the submission to be suspicious, but the forensic effort required to identify the source could be formidable.

For example, in one case a student had taken the completed essay of another from the in-tray on the instructor's desk, replaced the cover sheet with a new one and submitted the altered essay before the deadline. The dishonest student then sent a self-addressed envelope to the instructor asking for the return of the marked essay. The instructor was about to do this when two things happened. First, the office of the Dean of Studies queried an irregular change of grade for the student in a previous course, and secondly the student who had written the essay wrote to the instructor to ask why the essay had not been marked and returned. Forensic investigation by the instructor revealed that the writer of the essay had not only used the reference works listed in the bibliography but also had saved copies of all correspondence with government ministries who had supplied crucial resource material. Needless to say, the dishonest student had not used any of the resource material and had no record of any correspondence with the government ministries referred to in the essay. The primary evidence in the subsequent disciplinary proceedings was the essay with its false title page. This was kept in a safe during the proceedings to ensure that it did not mysteriously disappear. After a long drawn-out process involving the threat of legal action by the father, a prominent criminal lawyer, the dishonest student was expelled. It is rare that such clear-cut evidence is available. For example, "essay banks" may supply essays on disks. These would allow the clients to alter the essays on their own word processors to make them appear authentic.

A specific problem in essays is plagiarism, that is, the copying of published material. There can be great difficulty in detecting plagiarism and in applying appropriate penalties. One of the complications is that there is a hierarchy of offences ranging from failure to cite used reference material to blatant attempts to represent the work of others as one's own. Classicists point out that copying in Roman times was highly favoured as a sign of erudition. However, there was no attempt to deceive the listeners or readers as there is in the case of plagiarism, which of course occurs in many contexts such as the writing of books. In the assessment context, plagiarism is difficult to detect unless the examiners are aware of the wide range of commentaries and resource material. In one case an instructor accused a student of plagiarism because the expressed ideas and writing style were very mature. When the student responded with a fervent denial of plagiarism, the appeal board decided to refer the matter to a Shakespearian scholar. This person was able to find the plagiarised passage in one of Bradley's commentaries on Shakespeare's plays and so the instructor's intuition had been correct. This illustrates the problem that many examiners or instructors do not have the knowledge needed to identify plagiarised passages. At many institutions, there is a recurrent legend, which suggests that it may be apocryphal, of an instructor reading a student essay with both approval and an uneasy feeling of having read the same text many years ago. The story goes that the instructor finally identified the source of the plagiarised material as an essay on file in the library, written many years ago by the instructor when himself a student!

Failure to cite sources may result in lost marks, incorporation of plagiarised material may result in a zero grade for the essay and submission of a purchased essay may result in a failure on the course. The statistics of such forms of cheating are not generally available. Another offence which is rarely reported is the multiple use of an essay to satisfy the requirements, without permission, in more than one course simultaneously or to submit an essay used on a previous occasion for some other purpose. A much more serious offence is the purloining of another student's work so that the victim is regarded, initially if not finally, as the delinquent person.

Institutions also report that cheating in laboratory work is endemic. One way of avoiding this is to require all students to write laboratory reports under supervision before they leave the laboratory. The disadvantage to this is that the reports are incomplete and perfunctory without sufficient time for good analyses. Another stratagem is to have subtle differences in the apparatus at different lab stations so that identical results are transparently due to copying. Another technique is to change the laboratory conditions each time so that students using laboratory reports from previous occasions will reveal themselves by the use of obsolete data. Each laboratory demonstrator and instructor needs to have some forensic skills to detect cheating and copying. An encouragement to students to cheat is to require long, formal reports written outside the laboratory on experiments that have not changed for many years. Even normally honest students, overloaded with work in mid-term, may be tempted to cut corners in writing lab reports by making some use of predecessors' reports. One dramatic case involved the intimidation of a student by a fellow student who not only used the results of the other person but lost or destroyed the lab book of the innocent student. As this turned out to be but one of many acts of academic dishonesty, the offending student was eventually expelled. Several years later, there were enquiries from other institutions about the transcript of marks submitted by this same student applying for admission. It turned out that the transcript had been altered by changing F to B and by deletions of the academic penalties imposed. A few years later, *Time Magazine* carried an account of a student expelled from Patrice Lumumba University in the Soviet Union for both cheating and terrorising other foreign students. It was none other than the same student who had become *persona non grata* in the institutions of North America.

A form of cheating on marked tests is the plea that the marker neglected to assign marks to parts of the test. This would be a legitimate plea if the instructor had indeed failed to take note of complete or partial answers. However, there is also the possibility that the student adds material to the marked paper and then complains about an oversight. One instructor, alerted to this possibility and aware of a number of aggressively competitive students in the class, photocopied the marked papers before returning them. Sure enough, one dishonest student came back with the complaint that successfully completed answers had been overlooked. After allowing the student to commit perjury in addition to entering new material retroactively, the instructor devastated the student by revealing the photocopy of the original answer. The penalty in this case was that the student, a very bright candidate for medical school with an A grade average, was assigned a fail grade in the course by the head of the department.

Why do students cheat in the face of severe consequences? This is not a new problem. Indeed, the biographies of several famous people (it would be injudicious to mention names) state that expulsion from school or university on grounds of academic dishonesty was a lamentable aspect of the formative years of the person. In some ways, expulsion is a more serious penalty than a prison sentence. In the case of prison, one is said to have "paid one's debt to society" after the sentence has been served. Expulsion, on the other hand, may mean a lifelong stigma which precludes admittance to any bona fide institution of higher learning. There are many reasons for cheating such as fear of failure, ambitions for professional careers and so on. Weak students will cheat to avoid parental recriminations concerning failure or to save face within their peer groups. However, cheating is by no means confined to weak students and is often detected in its more pernicious forms among able and ambitious students who are upwardly mobile. It is often promising candidates for professional programmes such as education, law and medicine who cheat not only to enhance their own personal records but to devalue the accomplishments of their peers. Such students deserve little sympathy; they may be the criminals of the future. It is vital for the integrity of an institution and for the protection of honest students to have good forensic procedures to detect virulent forms of academic dishonesty.

Statistical data concerning the prevalence of various forms of cheating are hard to come by. While there is anecdotal evidence that plagiarism is widespread in essay assignments and that copying of laboratory results is endemic, departments and institutions keep the numbers to themselves. In the case of expulsions, institutions usually deal with these on a confidential basis. When dealing with serious forms of academic dishonesty, the governing body often goes into closed session with the exclusion of observers and reporters from the proceedings. Some institutions publish the sanctions that have been applied in the most serious cases, but it is rare for names and other forms of identification of individuals to be revealed. The most serious sanction is expulsion without the right of appeal and a permanent record of the offence on the official transcript. Some institutions allow an appeal after the lapse of a suitable period, say three years, with the possibility of erasing the record of the offence from the transcript after a period of blameless and satisfactory performance. An estimate of the numbers involved, based on a recent report to the Senate of Queen's University in Kingston, Ontario, would be of the order of five students in 10 000. A "Student Discipline Report" for a five-month period at the University of British Columbia (1994), which has 30 000 students, shows that there were 16 expulsions on grounds of academic dishonesty for the period, three for four months, eight for 12 months, four for 16 months and one for two years. There was a sensational case in the United States at Annapolis

Naval Academy, featured on a TV programme, in which there was wholesale cheating by a class of cadets. This was revealed because of the code of honour of the cadets which led many to confess their misdemeanours. Confession is all too rare, although it would be of benefit to all, especially the guilty parties, if confessions would reveal the truth of the matter.

One student who had already graduated was troubled by two occurrences of plagiarism during student days. Because of a profound religious experience, this student went to see the Dean of Studies to make a full confession and to request appropriate penalties. As it turned out, the student had been generally a good student and the plagiarism was of the type that would have enhanced grades rather than making the difference between pass and failure. As there was little that could be done about the official records retroactively, the Dean suggested as a penalty that the student take and pass two extra courses beyond the already fulfilled requirements for graduation. The student readily accepted this advice and the burden of guilt was somewhat assuaged by carrying out this assignment. Regrettably, this form of candour and confession is all too rare. Many former students may still carry with them a sense of guilt and shame because of concealed acts of academic dishonesty during their student days.

THE CRUCIAL ROLE OF BOARDS OF APPEAL

Appeal boards have delicate and demanding issues to resolve. Often it is the appeal of an aggrieved student against the treatment meted out by an instructor, a department or by an institution. A dependable set of appeal procedures and academic regulations is essential in making fair judgements in such cases. Certainly it is vital to seriously consider the grounds of the appeal and to investigate the case without prejudging the outcome. In the cases of charges of academic dishonesty against students, it is crucial to have unimpeachable forensic procedures that discriminate between unfounded allegations and hard evidence that will not crumble in cases of appeal to a higher authority or in the rare cases of legal actions in the courts. Being a member of such an appeal board is challenging and time-consuming, but the reward is to know that justice has been done in the end.

CONCLUSIONS

This chapter has argued that systematic evidence of cheating by students and unfairness on the part of instructors is difficult to obtain. Nevertheless, there is sufficient anecdotal and case history evidence to treat these threats to objective assessment seriously. In the light of this, institutions have a responsibility to provide adequate appeals procedures for students, and to monitor in a general way the conduct of examinations. Perhaps, above all, there is a responsibility to attempt to educate members of higher education institutions in the use of appropriate examination procedures.

NOTE ADDED BY EDITORS

Professor McLay has emphasised the necessity of having unimpeachable, or at any rate, the best possible forensic procedures to operate in the cases of charges of academic dishonesty against students. In fact, it is not unknown for a university's procedures to be under challenge in this respect, and regrettable when this occurs. An important case in England concerning disputed allegations of cheating in an examination is before the courts at the time of writing (late 1995): see for example *The Guardian*, London, 17 February 1990, p. 23. In another case, likewise involving a British university, two twin sisters were accused in 1987 of cheating and stripped of their first-class honours degrees. Four years later the university admitted it had been wrong and was determining how to redress the grievance: see *The Mail on Sunday*, London, 17 February 1991, p. 28.

CHAPTER 9

The Use of Assessment to Compare Institutions

John Gray

Homerton College, Cambridge, England

INTRODUCTION

The urge to compare the performance of educational institutions has been irresistible in most advanced industrial societies during the last decade of the twentieth century. Regardless of whether the concern has been to drive up "standards" or to inform parental choice, the consequences have been comparable around the world. And whether it has been the publication of information about secondary schools' performances in the GCSE public examinations in England, about French schools' success rates in the Baccalaureate or the graduation rates or test results of North American high schools, the issues and concerns that have emerged have been remarkably similar. How can one move beyond simplistic attempts to provide "raw" information on what the pupils in a particular school actually achieved to assessments of the effectiveness (or otherwise) of the institutions they have attended? How, in brief and in current parlance, can one determine the value which has been added to a pupil's performance by attending one institution as opposed to another?

This chapter is concerned to identify some of the problems to be faced in ensuring that schools are compared more fairly on a like-with-like basis. It is premised on the assumption that such comparisons will continue and that the most helpful service can be rendered by reminding those who would make judgements of some of the problems to be encountered.

THE POLITICS OF COMPARING INSTITUTIONAL PERFORMANCE: THE BRITISH CASE

In November 1992 the British government finally did what it had been threatening to do for a long time. It gathered together and then released summary

Assessment: Problems, Developments and Statistical Issues.
Edited by H. Goldstein and T. Lewis.

information about the examination results of every secondary school in England. Within a short space of time various "league tables" had appeared in the media. These rearranged schools' results (along one summary measure of performance or another) from the alphabetical order in which they had been released into rank orders, ranging from those with the highest scores to those with the lowest. A series of newspaper articles describing schools in terms such as "the best" and "the worst" accompanied their publication. Fierce debates about the appropriateness of the inferences being drawn about schools' performances followed on.

Comparisons of schools' performance have, of course, been undertaken since the latter half of the nineteenth century when Her Majesty's Inspectors of Schools were first recruited to assess the standards of education in the new system of mass state education (see Chapter 2). After a while, however, the system of "payment by results", which they were required to implement, fell into abeyance. For much of the twentieth century, while the performances of schools and other educational institutions were undoubtedly compared on an informal basis, a professional consensus emerged that such comparisons would not be undertaken in the public spotlight. Consequently the means by which such evaluations might be made remained largely underdeveloped.

It was in September 1978 that MP Rhodes Boyson, himself a former headteacher, broke the mould of professional assumptions. Comparing the results of a school in fairly middle-class and suburban Tameside with the results of an inner-city and severely disadvantaged comprehensive school in Manchester, he declared that the latter school (which he named) had the "worst results" in the city. A year later the newly returned Conservative government passed legislation requiring schools to publish their examination results in the prospectuses they handed out to prospective parents.

For much of the 1980s it was possible to compare the results of individual schools, although only in a very crude way. To do so, however, required considerable effort on the part of those so minded and there were few examples of individuals or organisations attempting to do so. With the passage of the 1988 Education Reform Act, however, all this changed. The new legislation envisaged that pupils would be tested at various so-called Key Stages (aged 7, 11, 14 and 16). Furthermore the results of individual schools would be made public. The main thrust, as with the earlier push for the publication of schools' exam results, was that parents should have access to what came to be known as the "raw" results. While the report of the Task Group on Assessment and Testing (TGAT) envisaged some arrangements for providing additional information about the social context of (and other pertinent circumstances facing) individual schools these were, in practice, rather modest and essentially of an advisory nature (Department of Education and Science, 1988).

A number of *ad hoc* strategies for releasing information began to be developed in different parts of the country before the turn of the decade. By the early 1990s, however, the government had decided that it needed to orchestrate a common national approach which culminated in the 1992 publication described above;

this approach was repeated (largely unaltered) in the two following years. In the run-up to publication, however, pressure for the government to take greater account of the circumstances facing individual schools began to mount. Around the time when publication was envisaged, there were signs of some concern that alternative approaches to judging the evidence on schools' performance might be contemplated. Such discussions began to coalesce around the idea that schools should be compared not only in terms of their "raw" results but also in terms of the "value" they had "added". This view eventually received official endorsement. Sir Ron Dearing's review of arrangements for the National Curriculum included a working party whose remit was to look at "value-added" issues. Their report, published in late 1994, was prefaced by some remarks from the Secretary of State for Education who, for the first time, "firmly committed (herself) to the development of robust national measures of the value-added by schools to children's education" (SCAA, 1994b). A series of pilot studies to support moves towards a national system were also commissioned.

THE NORTH AMERICAN EXPERIENCE OF ASSESSMENT-LED REFORMS

As the report *Testing in American Schools: Asking the Right Questions* (US Congress, 1992) makes clear, programmes of educational testing have been central to national debates in the United States about educational reform. Their role has been twofold; first, to document the need for change; and second, as "critical agents" within the reform process (Linn, 1993). In brief, the tradition of using standardised tests to compare the performance of schools, school districts and whole states has been well established.

Just how well-established standardised testing had become became clear during the so-called "Cannell Controversy" (Cannell, 1988). Cannell analysed the reports on pupils' performance from all 50 of the separate states and discovered, somewhat to his surprise, that all of them claimed that their pupils were performing at levels above what they deemed to be the "national average". Furthermore, up to 70% of all American pupils were being informed that they were performing at "above average" levels. Perhaps the most interesting feature of this debate, however, was the extent to which it turned out to be partially true. If states used outdated national norms, aligned their curricula to the tests being employed to evaluate them, chose their tests, in turn, to reflect their curricula, made it clear that test results mattered and were a "high stakes" issue, and engaged in a number of more or less ethical practices regarding "test preparation", then performance levels could be expected to rise (Linn *et al.*, 1990; Smith, 1991).

While individual states have adopted blanket testing strategies, there has been nothing on a co-ordinated scale comparable to that seen in Britain. However, the practice of using standardised test results to drive reform has continued into the 1990s with the further development of the National Assessment of Educational

Progress (NAEP), first started in the early 1970s (Mullis, 1994). The deliberate use of relatively small samples of pupils has mitigated some of the effects on individual institutions, although interest in "high-stakes" approaches to educational testing continues largely unabated. Meanwhile some states, such as Tennessee, have also explored the possibilities of using alternative approaches based on the more sophisticated procedures adopted by researchers of school effectiveness (Sanders and Horn, 1994).

THE TRADITION OF RESEARCH ON SCHOOL EFFECTIVENESS

As a direct consequence of these world-wide trends interest in the techniques routinely employed by researchers of school effectiveness has grown to unprecedented levels. Researchers have usually referred to estimating schools' levels of "effectiveness" but, in recent years, such terms have become overlain with nuances drawn from economic and business usage. The possibility of undertaking so-called "value-added" analyses owes much to the development in Britain, North America and elsewhere of a strong tradition of research on school effectiveness stretching back to the late 1960s and early 1970s.

Such research has three core assumptions. First, that schools should be compared on a like-with-like basis. Second, that it is the progress that pupils make from their respective starting-points that is the prime concern. And third, that educational institutions vary, to a greater or lesser extent, in their "effectiveness" in boosting their pupils' progress. Efforts to compare the performance of individual institutions, therefore, will usually be as firmly founded as the research traditions upon which they have been perched. The research provides some useful pointers to the possibilities of public comparisons between institutions, but also highlights some of the difficulties. However, the push for public information and accountability ignores the researchers' caveats at its peril.

This is not the place to summarise the long history of research on school effectiveness (for comprehensive summaries see, for example, contributions to Reynolds and Cuttance, 1992, Gray *et al.*, 1995 or Scheerens, 1992). It is sufficient to note its existence. What is of more concern here are the particular attempts to bend its methodologies to the problems posed by comparing institutions. Some of the very first efforts to apply such strategies are to be found in reports from the Inner London Education Authority during the late 1970s (see Byford and Mortimore, 1978). Several other local authorities were also beginning to experiment around this time (Gray, 1981). During the early-to-mid-1980s a number of them participated in the Contexts Project at Sheffield University funded by the then Social Science Research Council (see Gray *et al.*, 1986). While this project confined itself to the 11–16 age range, work at Newcastle University had begun to look at A-level results where information about performance in examinations at 16+ could be compared with subsequent performance two years later (see Fitz-Gibbon, 1985).

The issues surrounding comparisons between institutions attracted fresh interest among statisticians around this time. Their efforts combined to ensure that not only were the issues better understood but that more sophisticated models (and notably so-called "multilevel" ones) were in extensive use by the end of the decade (see Chapter 10). Goldstein (1984 and 1995) and Aitkin and Longford (1986) were prominent in this push and it was the latter whose pioneering study, using data from the Contexts Project, first fully alerted researchers of the extent to which different statistical models were capable of producing different estimates of schools' effects. Meanwhile, in the USA, Raudenbush and Bryk (1986), among others, had pioneered similar sophisticated statistical models of schools' effects which also underlined the importance of taking the "hierarchical" and "nested" structure of data on schools' performance more fully into account (see Bock, 1988 for a variety of world-wide contributions to these developments).

In brief, by the end of the 1980s, a clear sense had been achieved of both the theoretical and practical issues to be tackled in comparing the performance of different institutions. Many of the challenges of the 1990s have concerned the difficulties to be faced in translating strategies which have worked for researchers into "systems" which could be used by administrators and practitioners.

Since attempts to compare institutional performances are usually premised on the assumptions made by researchers of school effectiveness, it makes sense to consider some of the issues with which this group have had to deal before turning to some others which have emerged from the promulgation of such techniques in the arena of public evaluation. A number of different areas of concern may be identified.

THE CONSTRUCTION OF CONVINCING OUTCOME MEASURES

Several different approaches have been used to develop outcome measures for comparisons. At first glance the simplest has been to use "standardised tests" of one sort or another, such as the Scholastic Aptitude Tests developed in the USA by the Educational Testing Service to assist decisions about entry to higher education (see Chapter 11) or the tests of verbal and non-verbal reasoning, reading comprehension and mathematics originally developed by bodies like the National Foundation for Educational Research in Britain to inform selection decisions at the point of transfer from primary to secondary school. Such tests have typically been produced by testing agencies to possess many of the characteristics psychologists and statisticians have felt appropriate for ranking *individuals*. Different tests have been designed for different purposes (for a comprehensive review of many of the tests in current use see the latest edition of Buros, or Levy and Goldstein, 1984).

None the less, it has appeared a relatively easy matter to adapt them to the purposes of ranking *institutions*. This has, in some respects, been the case. Such tests have usually been developed to have highly acceptable "face" and other kinds of validity as well as reporting impressively high statistics regarding their "reliability". They have almost invariably also been deliberately constructed to have good statistical properties (such as discriminating well) which approximate to the assumptions that form the starting-point for most statistical modelling exercises.

Indeed, there has usually been only one flaw with such tests—albeit a rather fundamental one. Although they are often intended to be "curriculum-free" and to sample from the general domains of pupils' knowledge in a particular area or subject, pupils' performances on them are related, in part, to their previous encounters with the specific types of question they are asked to answer. Lower down the age-range, with tests of so-called "basic skills", this may be somewhat less of a problem, although even here it is difficult to avoid controversy. Higher up the age-range, as several international comparisons of performance have shown, the "opportunity to learn" various test items plays a significant part in pupils' measured performance (Walker, 1976; Keeves, 1992). Standardised tests, by their nature, fall foul of situations where there is less than universal agreement about the nature and foci of the curricula schools are supposed to be teaching their pupils.

Such concerns have contributed to considerable efforts being mounted to adopt and adapt the assessment measures actually being employed within the educational system; such measures have the considerable advantage that they are widely perceived as important and lay down (most of) what is to be taught to pupils (Madaus *et al.*, 1979). Initially, in Britain at least, this has meant focusing on the results of public examinations at 16+ and 18+. Until 1988 researchers were obliged to make only partially informed judgements about the equivalencies of grades in two separate examinations (the O level and the CSE); at the same time there has been the problem that not all pupils were necessarily presented for public examination. The population of candidates may have been smaller than the population of pupils in a particular institution (see Gray *et al.*, 1986). Since 1988, however, matters have eased a little in both respects, with the introduction of a single examination for all pupils—namely the GCSE—which has a single (eight point) grading scale and in which nearly all pupils are expected to take one or more subjects. None the less, some difficulties remain; the fact of different markers in different subjects making different assumptions counsels the need for caution in assuming strict equivalencies relating to grade assessments (Chapter 5).

The GCSE is an examination which pupils take at (or towards the end of) their period of formal compulsory schooling. The introduction of tests related to the National Curriculum in Britain represents one attempt to produce information at earlier stages of the pupils' careers. While there is much to commend in the various so-called Key Stage tests there is fairly widespread agreement that, in their present form, they leave something to be desired when it comes to comparing institutional performance (one of their intended purposes). There are several reasons for this, but the main one is that the descriptions of the different "levels" pupils attain are insufficiently differentiated. This is a position with which the School Curriculum and Assessment Authority (SCAA), the agency responsible for developing them, now seems to agree (SCAA, 1994b). At Key Stage 1, for example, almost the entire pupil population scores at one of just three levels with the majority being located (as the originators intended) at level 2. Although older pupils are distributed across more levels, the problem of a restricted range still largely remains. Another difficulty is that the results are not routinely standardised in terms of the ages of the pupils taking the tests, even though performance on them has been shown to be age related (Sharp *et al.*, 1994). An institution with more "older pupils" within the particular age range under scrutiny will appear to do better unless some account is taken of this factor (see Chapter 4).

There is one other area of debate. It is frequently assumed that different measures of schools' outcomes are highly correlated and that, consequently, it makes little difference which particular measure of outcomes is used to compare institutions. Research evidence from school effectiveness studies suggests that this conclusion needs to be qualified in two respects. First, different ways of "capturing" the same general areas of outcomes do lead to some differences in the institutions which are identified as more or less "effective"; the correlations between measures are usually on the high side (around 0.8 to 0.9) but not sufficiently high to suggest that they are completely interchangeable (see Sammons *et al.*, 1994 and Scheerens, 1992 for reviews). Second, correlations between different areas of outcomes are usually lower (typically between 0.3 and 0.7). The assumption that schools which perform well in one area will perform equally well in others is questionable. There are rather few examples of schools which are universally effective across outcomes.

It is easy enough to identify the various problems which attend the use of outcome measures as a means for comparing institutional performance. It is more difficult, in practice, to avoid all or most of them. Researchers and practitioners have usually felt constrained to work within the limitations of what has been readily available.

NON-COGNITIVE OUTCOMES

Almost all studies of school effectiveness to date have included outcome measures based on pupils' cognitive performance; a rather smaller number have looked at non-cognitive aspects. These latter studies are of interest both in their own right and in so far as they throw a different light on which schools come to be identified as the "more effective" ones. While there may be a general tendency for schools which do well in one area of pupil outcomes to do well in others at the same time, the relationships are, at best, merely trends (see Sammons *et al.*, 1996 for a review). One needs to be alert to the fact that schools may vary in their effectiveness according to the outcome of interest.

A variety of measures have featured in the research. The pioneering study of Coleman *et al.* (1966) used pupils' reported self-esteem; Epstein and McPartland

(1976) attempted to gauge the "quality of life" in North American schools; while Goodlad (1984) concentrated on aspects of pupils' satisfaction with their educational experiences. The British study known as *Fifteen Thousand Hours* (Rutter *et al.*, 1979) used measures of pupils' attendance and behaviour as well as their reported convictions for juvenile delinquency; a major study in Scotland employed measures of self-reported truancy alongside pupils' attitudes towards their school experiences (Gray *et al.*, 1983); while a study of inner London primary schools focused on a number of areas of pupils' attitudes and self-esteem (Mortimore *et al.*, 1988). More recently the National Commission on Education (1993) has recommended that the range of performance indicators considered alongside examination and test results should be considerably expanded to include levels of participation in post-compulsory education, attendance and drop-out rates as well as pupil and student attitudes.

While the attempt to broaden the range of outcomes under consideration has been generally welcomed, many of the same concerns attend their use as those relating to more cognitively oriented ones. How valid are the measures being employed? Do schools value higher performance on the outcomes involved? How reliable are they? These may vary. In the case of attendance measures there may be a high degree of reliability, although even here different approaches to data-gathering (registers versus self-reports for example) may yield different patterns; indeed, in the case of attitude measures, reported estimates of reliability tend to be low (often around 0.4) suggesting a high degree of variability between one occasion and another. And crucially, how can structures be built up which pay attention not only to the routine collection of such data, but also to their subsequent interpretation and utilisation? Most commentators agree that, desirable as their development is, such measures have yet to be taken as seriously as those measuring cognitive aspects of institutions' performance.

MEASURING PROGRESS AND IDENTIFYING MEASURES OF PRIOR ATTAINMENT

The simplest way theoretically to compare institutional performance would be to allocate pupils or students randomly to different institutions. This would ensure that whatever factors made a difference to their subsequent performance were equally distributed across the institutions. Of course such aspirations are, in practice, unrealistic. The researcher is faced with situations in which pupils have already made their choice of schools on a non-random basis. The challenge is to try to identify those salient characteristics (such as their prior attainments) which influence subsequent progress in relation to the outcomes of interest and to "take account" of them using the appropriate statistical procedures.

The ideal measure of "prior attainment" would be the pupils' score(s) on the same "test" or instrument being employed to compare outcomes. In practice, this approach is rarely feasible when periods of more than a year or two are being contemplated over which to monitor progress (see Chapter 5). Few tests are available which stretch sufficiently to cover longer periods of pupils' growth. In some subjects, such as mathematics, there is a fairly strong belief in pupils' following "developmental paths" of one kind or another and consequently, perhaps, a greater acceptance of the extent to which tests of different age-groups capture the same kinds of skills, although even these are vulnerable to the problem that different aspects of mathematics may be deemed appropriate (or inappropriate) for particular age-groups. Such assumptions become more strained, however, in subjects where the core nature of the subject is disputed. Tests of "English" of one kind or another may fall into this category. In some subjects, moreover, there may be some doubt whether suitable measures of prior attainment are available at all. Few schools (let alone school systems) collect information, for example, on pupils' early/prior attainments in French or other modern languages or about their general aptitudes for learning languages.

In brief, there are rather few circumstances in which the kinds of information researchers would ideally like to have available for the purposes of conducting comparisons can be obtained. It needs to be acknowledged that most of the procedures currently available are of the "second-best" type when it comes to the selection of measures of prior attainment. However, this general observation may be less true of the upper stages of secondary schooling where a certain uniformity of testing (or examining) procedures may begin to prevail.

The kinds of tests that are typically employed in school effectiveness research vary according to the age-groups being considered. For age-groups around the point of transition to secondary schooling, standardised tests of attainment in reading and mathematics which can be administered to whole classes at a time, produced by agencies such as the National Foundation for Educational Research (NFER) in England and the Educational Testing Service (ETS) in the USA, have been popular. In some areas tests of verbal and non-verbal reasoning have also been employed along with others which purport to measure pupils' "intelligence". Sometimes such tests have been deliberately put in place to facilitate comparisons between institutions higher up the system; more usually they have been administered for other purposes (such as screening for selection to more academic schools, for the identification of special needs or for the purposes of monitoring overall levels of performance in core subjects). However, they have only retrospectively been pressed into service for the purpose of making between-institution comparisons.

While it is difficult to be precise about such matters, the correlations between these various measures of prior attainment around the age of transfer and individual pupils' subsequent performances towards the end of compulsory schooling have ranged up to 0.7 (Gray *et al.*, 1990; Thomas and Mortimore, 1996). As it is the more sophisticated measures which have yielded the higher estimates, correlations of around 0.7 may serve as litmus-tests of the extent to which measures of prior attainment in any particular study of this age-range are likely to be credible. Slightly higher (but only *slightly* higher) multiple correlations have been reported when two or more measures of prior attainment have been available (see Willms, 1992; Thomas and Mortimore, 1996).

The ready availability of a wider range of tested and examined outcomes in the later stages of the secondary school has facilitated work on institutional comparisons. In England the various subjects of the GCSE have been available as subsequent predictors of performance in subjects taken two years later at A level; and in Scotland performance in the Standard Grade examinations, again taken at 16+, has been related to subsequent performance a year later in Highers. Correlations of around 0.6 to 0.7 have been obtained in this kind of work (Sammons *et al.*, 1994). In similar vein many school systems in North America have attempted to relate performance around this age to subsequent performance on graduation; the follow-up to the "High School and Beyond" study (Coleman *et al.*, 1983; Coleman and Hoffer, 1987) was a particularly prominent and early example of a national attempt to capitalise on this opportunity, although the tests used to establish performance levels and progress were (appropriately for a piece of research) of a rather abbreviated nature.

Efforts to establish comparable kinds of databases at the pre-school and early years stages to serve as measures of prior attainment for performance in the primary school have, to date, been somewhat less successful. There are a number of reasons for this including: the difficulty of obtaining any reliable measures of young children's performance without undertaking time-consuming and individual testing; debates about concepts such as "reading readiness" and their practical utility; a widespread awareness of the difficulties associated with the measurement of such concepts; the lack of quickly administered group measures of pupil attainment; and a climate of opinion among early educators that formal testing of young children was inappropriate and possibly harmful. In recent years, however, there have been signs of some movement away from such positions, driven in part by a concern that primary schools (like their secondary counterparts) might be unfairly judged, if data of a contextual or prior attainment nature were not available. At the time of writing, however, progress towards more wide-scale strategies has been patchy.

ALTERNATIVE MEASURES AS BACKGROUND CONTROLS

Measures of prior attainment have not always been readily available to researchers. On such occasions they have tended to use measures of pupils' socio-economic backgrounds as proxy variables for ensuring like-with-like comparisons between schools. The *Equality of Educational Opportunity Survey* conducted by James Coleman during the 1960s was among the first of a highly influential series of studies to use such a strategy (Coleman *et al.*, 1966). More recently a similar approach was employed in Teddlie and Stringfield's (1993) study of schools followed up in Louisiana over a period of 10 years. Schools systems often have some very rudimentary administrative information available about the incidence, for example, of free school meals. Researchers have usually found it expedient to supplement these with further evidence on parental background, occupation, education and other potentially relevant home circumstances. Gender and ethnic background have also been shown to be related to subsequent attainment levels and have consequently been included in the "controls" for comparisons between schools as well (see Gray *et al.*, 1990 for a summary).

It would be naive to argue that such background measures are as satisfactory for the purposes of controlling for relevant differences between schools as measures of prior attainment. In practice, however, the number of studies which have used the former almost certainly exceeds those which have relied on the latter. Furthermore, when several variables have been used to control for background circumstances, multiple correlations of around 0.4 to 0.5 have frequently been obtained (Gray *et al.*, 1990; Sulganik, 1994). While correlations of this order are a good deal lower than those which have been demonstrated using prior attainment measures, they are clearly a good deal better than having little or no information available at all.

THE DEVELOPMENT OF MORE SOPHISTICATED STATISTICAL PROCEDURES

Until the mid-1980s the statistical procedures researchers routinely employed to compare institutions' performances were relatively crude. In a typical analysis pupils' results would be aggregated to the level of the school and then various statistical analyses (mostly based on multiple regression procedures) would be performed on the resulting school-level averages. In a school system with 40 schools, therefore, the relationship between outcomes and intakes would be established on the basis of the so-called "line of best fit" covering the 40 (aggregated) data-points.

There had been growing awareness among statisticians that such an approach was unsatisfactory from the early 1970s onwards, but no clear strategies had emerged until the end of the decade. Around that time researchers were recommended to focus on the variations between pupils within individual schools at the same time. In a typical analysis using this approach the data for individual pupils in each school in the sample would be identified and a separate regression analysis run for each school in turn (see Burstein, 1980 for a summary). It rapidly became obvious, however, that the "line" for each school was liable to fluctuate considerably, not least because of the relatively small numbers of cases on which it might be based (say 50 to 100 pupils in a school). The search for a means of estimating the variation that occurred between schools in pupils' performance in comparison with the variation that occurred within them became paramount.

Some theoretical grasp on how to combine the two types of analysis began to emerge in the late 1970s and early 1980s, but the practical means of resolving the sizeable computing issues a typical data-set might involve was only discovered several years later. Three parallel but essentially equivalent approaches were developed (for further details see, for example, Aitkin and Longford, 1986; Goldstein, 1995; Raudenbush and Bryk, 1986).

A variety of terms have been used to describe such models, but in popular parlance the concept of a "multilevel" approach has probably come to dominate. This approach has influenced thinking about comparisons between institutions in several key respects as well as modifying some of the better-established conclusions about how schools differ in their effectiveness.

First, it has focused attention on the structure of variations both between and within institutions. The term "multilevel" is used to describe the "nested" structure of the data from which school effects are estimated. Pupils (referred to technically as level 1 units) are nested within schools (level 2), while schools, in turn, may be nested within organisational structures such as LEAs or state systems (level 3). Perhaps its biggest substantive contribution here has been to show that, while much of the public interest has been on differences in "effectiveness" *between* schools, much the larger part of the variation has lain *within* them. At the same time the approach has exposed the extent to which, within any hierarchically organised system of education, it has often been the lower-level units where much of the significant variation has turned out to be located. Thus differences between subject departments and individual classrooms have emerged as potentially much more important than differences, say, between local education authorities (Creemers, 1992; Willms, 1992). See also Chapter 10.

Many schools will, of course, be equally effective for all types of pupils. However, a second contribution of the multilevel approach has been to sensitise researchers to the possibility that institutions are "differentially effective". In other words, they may do better with some types of pupil as opposed to others —girls, for example, or pupils who were higher-attaining on entry (see Nuttall *et al.*, 1989 but also Jesson and Gray, 1991). Clear evidence for the existence of such effects should again counsel caution towards attempts to make simple summary judgements about schools' performances. For a sizeable minority of schools no such clear-cut judgements would appear to be possible.

Third, it has tended to discredit the idea that finely ranked "league tables" of schools' "effectiveness" are possible. The main substantive conclusion to be drawn from the analyses which have been conducted to date is that the considerable majority of schools achieve precisely the sort of results one would predict from knowledge of their intakes. A few may do substantially better while a similarly small number may do substantially worse. In brief, only broad-brush statements about "effectiveness" seem appropriate. These might take the form: "about as expected", "considerably better than expected" and "considerably worse than expected". The recent use of procedures for estimating the "uncertainty" attached to the estimates of "effectiveness" for specific schools has reinforced the view that a school's "effectiveness" is not a precisely estimable quantity (Goldstein and Healy, 1995). The likelihood that a different sample of pupils from the same school would yield a similar estimate of its apparent "effectiveness" can be calculated and a "band of 'uncertainty'" constructed about the resulting estimate. Where such "bands" have been constructed they tend to show that the performances of some two-thirds to three-quarters of schools cannot properly be distinguished from each other. Again caution in interpreting results is called for.

CONCLUDING OBSERVATION

The main conclusion to be drawn from research on school effectiveness to date is that there are a number of pitfalls to be encountered when comparing the performances of individual institutions. Many of these are well understood; some, however, are not. A considerable measure of caution should accompany any interest in such comparisons.

ACKNOWLEDGEMENT

I should like to thank my colleague Rex Watson for his assistance with certain statistical matters.

CHAPTER 10

The Statistical Analysis of Institution-based Data

Geoffrey Woodhouse and Harvey Goldstein

Institute of Education, London, England

INTRODUCTION

In addition to its uses for promoting learning, for understanding the factors associated with learning and for certification of individuals, assessment is being used increasingly to make comparisons between institutions. Perhaps the most common example is that of schools, where in many countries aggregated examination results or test scores or even teacher judgements are used to produce rank orderings, sometimes referred to as "league tables" using a sporting analogy. Such lists, often informally compiled, are then often used to promote schools with high scores and in some cases, as in parts of the United States and in England, have become part of an official "accountability" mechanism (Goldstein, 1993). In addition, the comparison of subunits such as subject departments has been advocated and implemented (Fitz-Gibbon, 1991). The extension of these ideas to other educational institutions, including universities, has also been discussed. Since a principal aim of educational institutions is to promote learning it would appear natural to evaluate their accomplishment of this aim by comparing the performances of the students who attend them. Of course, educational institutions also have other objectives by which they should be evaluated, but that is not a principal concern of this chapter, even though many of the problems will be common.

There is an important distinction between using measures of student achievement to study factors associated with that achievement which may also explain institutional differences, and the use of student achievement measures to rank institutions for purposes of public or private (within-institution) accountability. The former may well provide information important for implementing the latter (for example regarding the influence of social background on achievement), but is essentially a matter for detailed and careful research, often known as "school effectiveness" research. In the latter case, and based on available knowledge, the intention is to discriminate among institutions and to decide when observed

Assessment: Problems, Developments and Statistical Issues.
Edited by H. Goldstein and T. Lewis.

institutional differences are both statistically and substantively significant. We shall refer to schools throughout most of this discussion since most attention has been directed towards these.

CONTEXTUALISATION

In evaluating any institution in terms of outcome measures, such as student examination results, it is clear that contextualisation is important. By this we mean the taking account of factors which influence the outcome measures, and over which the institution itself has little control. Thus, examination results at the end of a period of schooling typically are well predicted, with correlations sometimes as high as 0.7, by the achievements of the students measured at entry to that school. Likewise, ethnic background and poverty are predictive of examination performance, even after allowing for initial achievement (Nuttall *et al.*, 1989). It is also generally true that schools differ, often considerably, in terms of the initial achievements of their students, their ethnic backgrounds, etc. In the interests of equity, therefore, comparisons between schools should make adequate adjustments for such factors. In other words, schools and institutions in general should be held accountable for the things which they can be expected to influence, rather than for the characteristics the students bring with them when they enter.*

This principle underpins attempts to devise measures of "value added" against which to judge schools. Some light may be shed on the difficulties involved by studying the statistical methods which have been used in school effectiveness research, their data requirements, some of their findings, and some of their limitations.

STATISTICAL MODELLING

If in judging schools by their outcomes we are to make adequate adjustments for factors beyond the schools' control, we must propose models for the ways in which those factors influence the outcomes, and test the models. This process can be illustrated by the following simple example.

Suppose we have data on each student in a given cohort in a given school (A), in the form of a score obtained on intake (at age 11, say) and an outcome score on leaving (at age 16). Ignoring for the moment any concerns about the validity or reliability of those scores we may plot one against the other as in Figure 10.1.

We note a general tendency for students who score highly on intake to score highly on leaving. This tendency may be modelled in a variety of ways, the

* It could be argued that institutions can influence the kinds of students they admit, by processes of selection, and that therefore intake achievement is a legitimate basis for comparison. This, however, has little to do with the effect that the institution has on the students once they arrive and it is this aspect with which we are concerned.

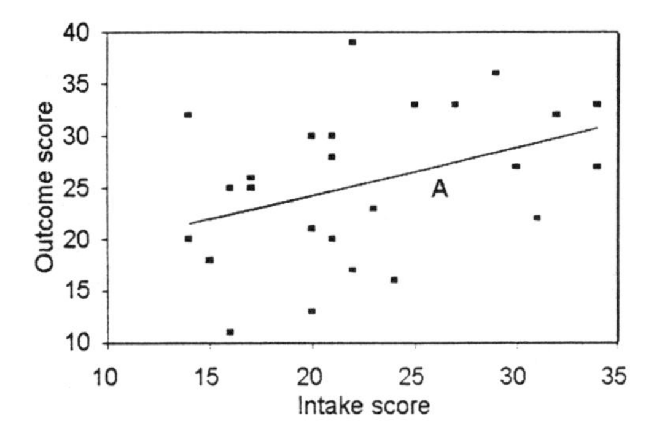

Figure 10.1 Intake and outcome scores for school A

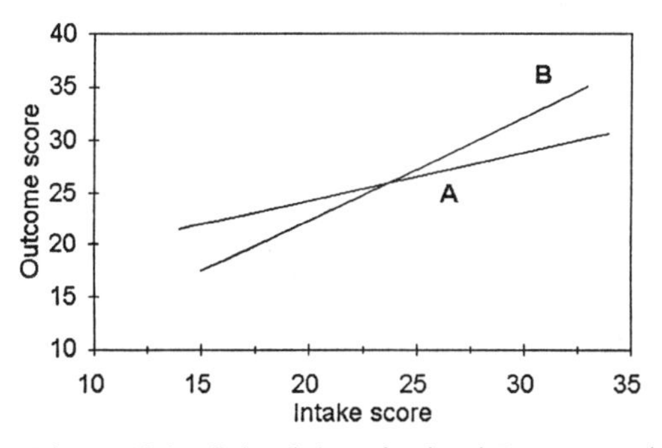

Figure 10.2 School A and school B compared

simplest being the straight line which has been superimposed on the plot of the individual measurements in Figure 10.1. Such a summary line may be estimated by the technique known as linear regression, and a point on it may be interpreted as giving the expected outcome score of a student with a given intake score, according to that model.

Suppose now that we have a second school (B), with similar data. Using the same procedure as before we may obtain a summary line for comparison with school A, as in Figure 10.2.

A number of observations can be made. The first is that, even with this very simplified model, it is not possible to say simply which school is the more "effective". Low-achieving students at intake can expect, on average, to achieve higher outcome scores in school A, whereas the reverse is true for those students with high intake achievements. Often referred to as "differential effectiveness", such patterns are common and illustrate that institutional comparisons may be complex and multidimensional.

Secondly, some uncertainty attaches to the estimates of the lines, even if we assume the model is correct. Subject to further assumptions about the scores themselves, the extent of this uncertainty can be estimated, and it may be that as a result the two schools cannot be separated one from another. It is worth mentioning here that school average scores (the so-called raw scores used in many published league tables) are derived from an even simpler model (a horizontal line for each school) and are subject to uncertainty in the same sense.

Thirdly, it is uncertain that the model is correct. We have deliberately simplified and it is indeed unlikely that such a simplified model is correct. But even with much more information about schools and students it remains necessary to check the sensitivity of findings to changes in the model used.

Fourthly, all assessments are subject to measurement error (see Chapter 4). Proper adjustment for such errors generally changes the estimates and may lead to a change in the model.

The above observations apply to most statistical modelling. The interpretation of the results may be subject to further problems, for example in the present case to changes over time in assessment instruments and in schools. We return to this issue later.

AGGREGATE-LEVEL MODELS

The example in the previous section assumed the availability of data on each individual student. In some cases, however, we may have only aggregated data in the form of school mean intake and outcome scores, proportions of students entitled to free school meals, etc. In such cases usually only very limited kinds of judgement can be made.

Consider again the simple case where we wish to model outcome scores as a function of intake scores only. Figure 10.3 illustrates how this might be attempted using aggregated scores only.

Each point plotted in Figure 10.3 corresponds to a different school, its position being determined by that school's mean intake and outcome scores. The straight line now represents a relationship which is assumed to exist for all schools between their mean intake and outcome scores: a point on the line indicates the mean outcome score which a school with a given mean intake score would be expected to achieve, on the assumption of that relationship.

Such a model has a number of deficiencies. First, it says nothing about the differential effectiveness of schools. The slope of the line does not show, for any school, how the outcome scores of its individual pupils might be expected to depend on their intake scores. Secondly, the use of such a model to compare schools amounts to comparing how far their respective points are displaced above or below the line. These displacements, or residuals as they are called, clearly depend on the model which is assumed. With relatively few data points (only one per school) it is quite possible to produce alternative models which fit the data equally well, but which produce quite different rankings among the residuals.

Woodhouse and Goldstein (1988) illustrate empirically how comparisons

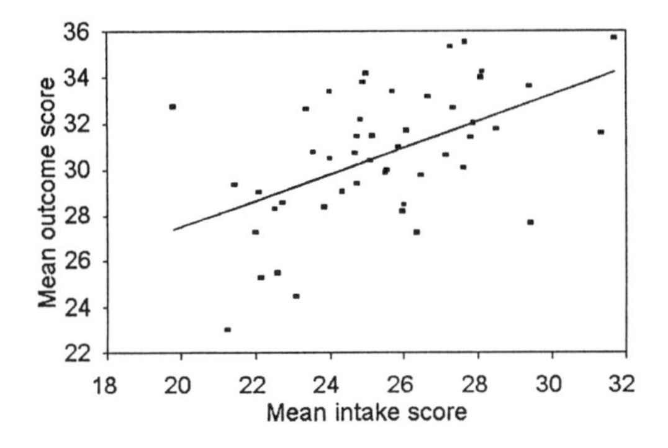

Figure 10.3 Mean intake and outcome scores for 48 schools

based upon such models are unstable. As the number of predictor variables increases so the instability increases, and an aggregate-level model becomes less and less useful. A further discussion is given by Goldstein (1995).

INDIVIDUAL-LEVEL MODELS

We have seen that, although our aim is to produce school-level estimates of effectiveness, we need individual-level data to do this. Consider again the simple model illustrated in Figures 10.1 and 10.2. We may write it as follows:

$$y_{ij} = a_j + b_j x_{ij} + e_{ij} \tag{10.1}$$

where y_{ij} is the outcome score for student i in school j, and x_{ij} is the intake score for that student. Each school has its own intercept a_j and slope b_j, and e_{ij} is the displacement of student i's outcome score above or below the score predicted for that student by the straight line for school j.

If we have a collection of schools which we wish to compare, and intake and outcome data for a sample of students, we can proceed to fit the model to our data, obtain estimates of the a_j and b_j, and use these together with estimates of their standard errors (representing sampling and year-to-year variation) to make judgements about differences. The traditional procedure is to estimate the a_j and b_j as separate parameters. Suppose we have 200 schools. This will give 200 intercept parameters a_j and 200 slope parameters b_j. If some of the schools have small numbers of students then the corresponding estimates will have very large standard errors, and some of the estimates themselves will have very large or very small values which may be quite different from their "true" values.

Now equation (10.1) represents a particularly simple model. If we wish to make properly contextualised comparisons, we must try to ensure that we take account of as many relevant influential factors as possible. In addition to adjusting for intake score we may also need to take account of social, ethnic, gender and other factors. There may also be interactions between factors so that, for example, gender differences may depend upon the intake score. The composition of the student body also may be important so that, for example, for particular intake scores the expected outcome score may depend on the average achievement of the other students in the school (Steedman, 1980). Furthermore, schools may vary in the extent to which such interaction and compositional effects are important, just as the effect of intake score varies between the two schools in Figure 10.2.

If our model is misspecified by omitting an important factor, or by modelling it inadequately (for example by fitting a straight line for prior achievement when the relationship is markedly non-linear), then any inferences about institutional differences will be suspect. But each factor which we add to equation (10.1) and which varies in its effect across schools will add 200 to the number of parameters which we need to estimate. As the number of schools increases so does the number of parameters and the computations quickly become unmanageable. For this and other reasons which we shall outline, many data analysts prefer to treat such effects as random variables, and this leads them to use multilevel models.

MULTILEVEL MODELS

Consider again equation (10.1):

$$y_{ij} = a_j + b_j x_{ij} + e_{ij}$$

and write, $a_j = a + u_j$, $b_j = b + v_j$. Equation (10.1), rearranged, becomes:

$$y_{ij} = a + bx_{ij} + u_j + v_j x_{ij} + e_{ij} \quad (10.2)$$

If, now, we regard $E(y_{ij}) = a + bx_{ij}$ as the population expected relationship and u_j, v_j as random variables, each with mean zero, an associated variance (between schools), and a covariance one with another, then equation (10.2) becomes what is known as a multilevel model, in this case a two-level model. It is so called because there are two levels at which random variation exists: level 1 (the student level), with random variable e_{ij}, and level 2 (the school level), with random variables u_j, v_j.

This formulation of the model has several benefits. The first and most obvious is that the number of parameters to be estimated is kept reasonable. Considered as a multilevel model, equation (10.2) requires the estimation of only six parameters: the two fixed parameters a and b, the variances of u_j, v_j and their covariance, and the variance of e_{ij}. Each additional factor with an effect which varies across schools adds *at most* $2 + r$ to the existing number of parameters in the model, where r is the existing number of level-2 random variables (currently two). This number does not depend on the number of schools represented in the data.

This feature allows us to explore more realistic models. Modelling variation between schools in this way is efficient both statistically and in the sense that it allows the analyst to focus attention on the effects that do vary significantly between schools, provided always that adequate data are available. Thus Nuttall *et al.* (1989), using this method, found that ethnic-group differences varied across schools, as did gender differences. Goldstein and Thomas (1995) studied the effects of GCSE examination scores (obtained at age 16 by students in English and Welsh secondary schools) on subsequent results at A level (taken two years later) by exploring different groupings of the candidates according to their GCSE scores. Institutions were found to be differentially effective for different groups of candidates, and by limiting the number of GCSE groups to three it was possible to present these differences in a comprehensible way.

Technically, this is done by using "posterior" or "predicted" estimates of the level-2 (institution-level) residuals. For equation (10.2) these are written $\hat{u}_j$, $\hat{v}_j$, and their values are based upon the model estimates of the variances and covariance of u_j, v_j. Also known as "shrunken" estimates, the mean value of any one of them for a given number of students in a school will be closer to the overall population mean of zero than the true unknown value. This shrinkage factor decreases as the number of students per school increases, and its effect is to avoid very extreme estimates arising as a result of small numbers with consequently large sampling errors. In other words, where the number of students in a school is small there is little information available about that institution and it seems reasonable that the corresponding estimates should be placed close to the overall mean. These shrunken estimates themselves have standard errors and these can be estimated.

THE INTERPRETATION OF RESULTS

We mentioned earlier that, in addition to the usual caveats that must apply to any statistical analysis, the comparison of institutions by means of outcome measures is subject to further problems of interpretation. Before we discuss these we review the matter of statistical uncertainty, and show how results may be presented in such a way as to take account of this.

Consider a simple two-level model with only one random variable at level 2, for example:

$$y_{ij} = a + bx_{ij} + u_j + e_{ij} \tag{10.3}$$

By contrast with equation (10.2), this model assumes that the relationship between outcome and intake score varies between schools only in respect of the intercept. Such a model (with additional non-random terms) might be fitted to one GCSE group in the analysis of A-level results referred to in the previous section. In fact, the data for the present illustration come from an analysis of 11-year Mathematics test scores in one English LEA, with adjustment made for intake achievement at the start of junior school (that is, at age 7). A full description of the analysis is given by Goldstein (1995).

Having estimated the parameters of the model we may obtain estimates of the shrunken residuals $\hat{u}_j$, together with their standard errors. These estimates may be illustrated as in Figure 10.4. The bars around the estimates represent "uncertainty intervals" which are constructed from the standard errors of the $\hat{u}_j$ and are

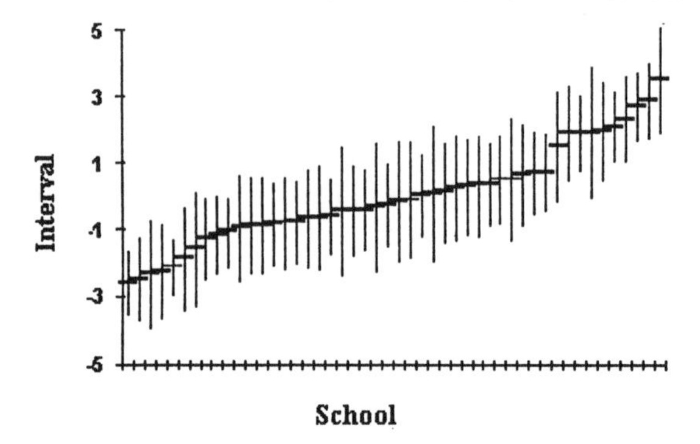

Figure 10.4 Uncertainty intervals for shrunken junior school residual estimates

interpreted as follows. Given any pair of schools in the sample, their values $\hat{u}_j$ are judged to be significantly different at the 5% level if and only if their corresponding intervals do not overlap. Goldstein and Healy (1995) give a detailed justification for the use of this procedure.

It can be seen that about four-fifths of all possible pairwise comparisons do not allow any separation of schools in the above sense. It appears, therefore, that while a ranking of schools in this way can be useful as one piece of evidence about institutional performance, it cannot be used by itself to provide definitive statements about differences for most institutions being compared. In some cases such displays may be useful as screening devices to identify extreme institutions for further study, but they do require careful handling to ensure that apparent differences are not misinterpreted.

We now turn to other difficulties that attend the use of outcome measures for institutional comparisons. First, it is important to ensure that the intake and outcome measures really are comparable across institutions. This will normally be the case when exactly the same measuring instruments are used, assuming that these are felt to be equally appropriate among institutions. Likewise, where a proper equating has taken place (see Chapters 4 and 5) we may be satisfied that comparisons are justified. In other cases, however, there may be some doubt, so that results of analyses may be difficult to interpret. This might occur if each institution were allowed to select its own version of the outcome measure, as occurs in the school-leaving examination system (GCSE) in England and Wales, or where an uncontrolled use of item banks was involved (see Chapter 4). The problem is particularly acute in the case of higher education institutions where there is often little formal attempt to ensure comparability of outcome certification measures.

Second is the problem of historical relevance. Because of the need to contextualise outcomes by using intake measures, the information about institutional differences is inherently outdated. To see this, suppose that we wish to judge secondary schools in terms of examination scores of students when they leave at the age of 16 years, using intake data at 11 years, that is five years earlier.

By the time results become available they will refer to a cohort of students who began their progress through the schools at least six years before the next current cohort is due to start. The use of such historically bound information for purposes of school choice, or even for resource allocation, will be problematical given that many schools can be expected to have changed noticeably during the intervening period.

The problems of comparability and historical relevance reinforce one another if part of the purpose of comparing institutions is to find out how they change over time. Because choice of tests, both at intake and on leaving, may change from year to year, comparability over time cannot be assumed. If, however, interest centres on comparisons of progress made over a single year, then the problem may not be so serious. Thus, we may well wish to compare institutions in terms of the progress made during the first, second, etc. years, and in view of the fact that students generally change teachers from one year to the next, this may constitute a more useful series of questions to ask. It does, however, require collection of more intensive data than merely information at intake and for outcome for a single phase of education.

Institutions have been found to be differentially effective for different kinds of student: they are also differentially effective for different kinds of outcome. Thus, for example, in a study of secondary-school outcomes in a single LEA in England, Thomas and Mortimore (1996) found low correlation (less than 0.5) across schools between performance in English and in Mathematics, after adjusting for intake, and only moderate correlation (less than 0.7) between performance in either of these and total outcome score. Such totals, or equivalently proportions of students obtaining set numbers of grades, particularly when aggregated over collections of subjects which vary widely between students and between schools, would seem to be less useful for institutional comparisons than more well-defined outcomes.

CONCLUSIONS

We have seen that aggregated data alone are not sufficient to enable institutions to be compared. The use of individual-level data, moreover, demonstrates that institutions may be differentially effective for different kinds of individual as well as for different outcomes. To search for some simple measure of "value added", to be used to rank institutions in place of unadjusted data, would be misguided. School effectiveness research, using individual-level data which are not yet routinely available, has revealed complex patterns of variation between institutions, but has also shown that the majority of institutions in any one study cannot be separated statistically one from another on the basis of their results. The number of students upon whom data are collected in any one institution typically is not large, and this implies large uncertainty intervals which do not enable sufficiently precise separation among the institutions. This problem can be expected to become more severe if we focus attention on specific outcomes, for example the results in individual subjects in secondary schools, when these involve still fewer students. In addition there is the problem of making adequate adjustment for measurement unreliability.

There are further problems associated with using assessment data for the purpose of individual institutional comparisons. The data need to be contextualised properly, but that will often lead to diminished relevance for current decision-making processes. In fact, the situation in this respect may be even worse than has already been suggested since there is some evidence (Goldstein, 1995; Goldstein and Sammons, 1995) that merely adjusting for intake achievement at the start of a phase of schooling may be inadequate: achievement and institutional membership prior to that point also may have important influences.

This chapter has focused on the problems inherent in attempts to compare specific institutions on the basis of their outcomes. It should be emphasised that research into the factors associated with institutional differences, using multilevel modelling techniques, is a valuable activity that can help us to understand what it is that promotes or hinders student progress.

CHAPTER 11

The Extent and Growth of Educational Testing in the United States: 1956–1994

George F. Madaus and Anastasia E. Raczek

Boston College, Chestnut Hill, Massachusetts, USA

INTRODUCTION

Test use in United States (US) schools has increased enormously over the last few decades. Currently, tests are being used for a multitude of purposes including: certification or recertification of teachers, promotion of students from one grade to the next, award of diplomas at the end of secondary schooling, assignment of students to remedial classes, allocation of funds to schools and school districts, award of merit pay to teachers on the basis of their students' test performance, certification or recertification of schools and districts and, in at least one instance, placement of a school system into "educational receivership". More generally, tests are now widely being touted as instruments of national education reform and renewal.* In short, tests are increasingly used for high stakes decisions. They might arguably be called the coin of the educational realm (Haney *et al.*, 1993). This chapter documents the extent and growth of educational testing in the United States in recent years.†

We begin with a description of the growth and extent of educational testing in the United States using a number of direct and indirect indicators. Second, we describe reasons for this growth. Third, we offer estimates of the costs associated with educational testing. Finally, we describe current trends that will shape the future growth of elementary and secondary school testing.

Two points need to be made at the outset for readers not familiar with the

---

* For claims about the value of a national test see, for example, US Department of Education (1991).

† Here we concentrate only on educational testing. The material in this chapter is largely drawn from Haney *et al.* (1993), which contains a full description of the testing market in the US including: licensing and certification tests for a wide range of blue-collar and professional occupations, military testing, industrial testing, psychological testing and various spin-off industries associated with testing.

Assessment: Problems, Developments and Statistical Issues.
Edited by H. Goldstein and T. Lewis.

American testing scene. First, testing in the United States is a very big, very lucrative, unregulated industry. Haney *et al.* (1993) estimate that the revenues from the sales of elementary and high school (secondary) tests and related services, such as scoring and reporting (called the Elhi market), is estimated to be between one-half and three-quarters of a billion dollars annually. To appreciate educational testing in the US therefore, one must clearly keep in mind that the institution of testing is at its root commercial and competitive.

Secondly, European readers need to appreciate the geographic size, population and student demographics of the United States system. The US educational system has a long tradition of local control. For example, the federal government contributes only about 6% of the funds spent on public schools (National Center for Education Statistics, 1994) and there is currently no national curriculum or assessment; given the Republican success in the 1994 elections, there is little likelihood this will change in the foreseeable future. In fact, there is serious consideration being given to closing the federal Department of Education and merging its functions with the Commerce Department. Responsibility for education generally falls to each of the 50 states and to the school district (LEA) in which a school resides. However, the extent to which state or district standards influence activity within individual classrooms varies greatly from region to region. Some states, like New York and Kentucky, have strong state Departments of Education, while others, such as Massachusetts and New Hampshire, have relatively weak ones. While space does not permit a detailed description of the US Elhi system, Table 11.1 attempts to put it in perspective by comparing Western European countries to the United States along the dimensions of size, population and student enrolments (Madaus and Kellaghan, 1994).

Examination of Table 11.1 shows that the majority of European countries—nine in all—are equivalent to a single American state across all three dimensions. For example, Belgium is equivalent to Maryland in size, and to Michigan in enrolment and population. Table 11.1 also reveals that the "large" European countries—France, Germany, Italy, Spain, and the United Kingdom—are relatively small compared to the US in size and roughly equivalent to only three or four of the largest American states in terms of population and enrolment. For example, while about the size of Oregon, the United Kingdom's population is roughly equivalent to that of California, New York, and Ohio, and its student enrolment compares to that of California, Pennsylvania, New Jersey and Minnesota.

THE GROWTH AND EXTENT OF EDUCATIONAL TESTING

While standardised commercial testing has been part of the American scene since the 1920s, the last two decades have seen a vigorous growth market for testing, with no decline in sight. In order to sketch the dimensions of the educational testing market place we first offer direct estimates of the extent of testing and then present indirect indicators on the extent and growth of the market.

DIRECT INDICATORS OF GROWTH

Since most students take more than one standardised test in a given year, one must distinguish clearly between the number of tests administered and the number of people tested annually. Because much of the source data on the extent of testing reports numbers of people tested in specific kinds of testing programmes, Haney *et al.* (1993) used an estimate of the number of tests taken by each person in each testing programme. For example, a student may sit for a basic skills test battery that consists of three separate tests, i.e. reading, mathematics, and listening skills. Since she/he receives a separate score for each portion of the battery, Haney *et al.* considered each separately scorable portion as a separate test in estimating the number of tests given annually.

Using this rubric, Haney *et al.* estimated that the number of standardised educational tests administered annually in state-mandated testing programmes, school district testing programmes, tests administered to special populations, and college admission (third level) testing programmes to be between 140 million and 400 million tests each year (see Table 11.2). This estimate represents the equivalent of between three and nine standardised tests administered annually to each of the nation's roughly 44 million students enrolled in public and private elementary and secondary schools.

Table 11.2 also reveals that local educational authorities (LEAs) (school districts) account for the largest testing component closely followed by state mandated programmes. LEAs generally purchase tests and related scoring services from one of three large testing companies: CTB/McGraw Hill; The Psychological Corporation (PsychCorp); or Riverside, a subsidiary of Houghton Mifflin (HM). In the past, states often used an off-the-shelf battery from one of the main publishers, but now state testing programmes are more likely to establish a contract with a company to build a test battery to specifications. Two populations of examinees, special education students and bilingual students, are tested extensively; primarily this is due to federal mandates described below. Third level (higher education) admission testing is dominated by two companies, Educational Testing Service (ETS) and the American College Testing Program (ACT).

Indirect indicators of growth

Reliable longitudinal data on the number of tests administered does not exist. Nevertheless, three indirect indicators show growth in the volume of testing over time: recent increases in the numbers of state-mandated testing programmes, sales figures from test publishers and references to testing in the education literature (Haney *et al.*, 1993).*

* Space does not permit us to describe a fourth indirect indicator used by Haney *et al.* (1993), that is, commentary from analysts of the stock market which describe testing as a growth industry worthy of investment.

Table 11.1. Comparative population and enrolment of western European countries and the United States

Country	Population 1991 Europe	1990 Comparable US		Size (sq. mi.) Europe	Comparable US		Enrolment Europe*	1989 US Public†	
Austria	7 700 000	New Jersey	7 703 188	32 375	VA	39 704	1 125 024	GA	1 126 535
Belgium	9 900 000	Michigan	9 295 297	11 781	MD	9 837	1 561 783	MI	1 576 785
Denmark	5 100 000	Missouri	5 117 073	16 631	$\frac{1}{2}$ ME	30 995	2 854 682	NY	2 565 841
Finland	5 000 000	Maryland	4 781 468	130 119	GA &	58 910	810 952	WA	810 232
					ND	70 665			
France	56 700 000	California	29 760 021	211 208	80% TX	262 017	9 465 580	CA X 2	4 771 978
		New York &	17 990 455						
		Ohio	10 855 000						
Germany	78 700 000	California		137 838	MT	145 388	11 291 788	TX × 2	3 328 514
		New York						CA	4 771 978
		Ohio &							
		Missouri							
Greece	10 100 000	Ohio	10 847 115	50 961	AL	50 767	1 708 355	OH	1 764 410
Ireland	3 600 000	Arkansas	3 665 228	27 136	WV	24 282	867 035	MA	825 588
Italy	57 700 000	California,	29 760 021	116 500	AZ	113 508	8 182 813	CA	4 771 978
		New York	17 990 455					NY &	2 565 841
		& Ohio	10 855 000					MA	825 588
Luxembourg	400 000	Wyoming	453 588	999	$\frac{1}{2}$ DE	1 932	45 413	$\frac{1}{2}$ WY	97 172
Netherlands	15 000 000	2 × New		16 041	VT and HI	9 273	2 752 000	NY and RI	2 565 841
		Jersey				6 425			135 729

Norway	4 300 000	Louisiana	4 219 973	125 049	NM	121 335	683 990	MD	698 806
Portugal	10 400 000	Ohio	10 847 115	35 550	IN	35 932	1 781 060	FL	1 789 925
Spain	39 000 000	California	29 760 021	194 884	CA &	156 299	7 484 487	CA	4 771 978
		& Ohio	10 847 115		TN	41 155		PA &	1 674 161
								NJ	1 076 005
Sweden	8 600 000	2 ×		173 800	IL &	55 645	1 174 170	NC &	1 080 744
		Louisiana			IN &	35 932		DE	97 808
					UT	82 073			
Switzerland	6 800 000	Georgia	6 478 211	15 941	2 × MA	7 824	1 005 175	NJ	1 076 005
UK	57 500 000	California		94 247	OR	96 184	7 691 792	CA	4 771 978
		New York &					(Eng/Wales)	PA &	1 674 161
		Ohio					743 986	NJ	1 076 005
							(Scot.)	MN	739 553
TOTAL	377 500 000		248 709 873	1 391 060		3 540 939	61 230 085		40 542 707

* Figures based on different years and include primary, secondary, vocational and special education, public and private institutions.
† Figures for some states may include pre-kindergarten.
Source: Kellaghan and Madaus (1995).

Table 11.2 Numbers of Educational Tests Given Annually in Late 1980s

State mandated testing programmes	Low estimate	High estimate
Number of students tested	11 000 000	14 300 000
Number of tests per student	3	5
Subtotal	33 000 000	71 500 000
School district testing programmes		
Number of achievement tests	82 668 966	248 006 898
Number of ability tests	2 952 463	23 619 705
Subtotal	85 621 429	271 626 602
Special populations		
Number of tests for special education students	7 920 000	19 800 000
Number of tests given to bilingual students	3 600 000	10 800 000
Subtotal	11 520 000	30 600 000
College admissions testing		
Subtotal	13 034 318	21 759 548
TOTAL	143 175 747	395 486 150

Sources: Haney *et al*., 1993; Office of Technology Assessment (1987).

State-mandated programmes

Figure 11.1 shows the growth in numbers of state minimum-competency and assessment programmes from 1960 through 1985. There was a steady rise in numbers of state-mandated assessment programmes, from one in 1960 to 32 by 1985. The rise in numbers of minimum competency testing programmes at the state level was even more dramatic. The curve rises sharply from one such programme in 1972 to 34 by 1985. Naturally, with every state mandate, the number of students tested—and hence the number of tests administered—increased. By 1985, state testing programmes were the most important market for publishers' group achievement and ability tests (Fremer, 1989). Although we were not able to update Figure 11.1 to 1994, a 1987 report noted that 43 states engaged in some form of state-mandated achievement or competency testing (OERI State Accountability Study Group, 1987).

Sales of tests

Though comprehensive data on the dollar volume of test sales are not available—most publishers consider such information proprietary—data on revenue figures are available for four of the largest players in the educational testing market: ETS between 1970 and 1991, National Computer Systems (NCS) between 1980 and 1990, ACT between 1972 and 1991 and Scantron between 1980 and 1989*. ETS and ACT control the college entrance exam market. ETS is also

* NCS and Scantron are in the test scoring and reporting business.

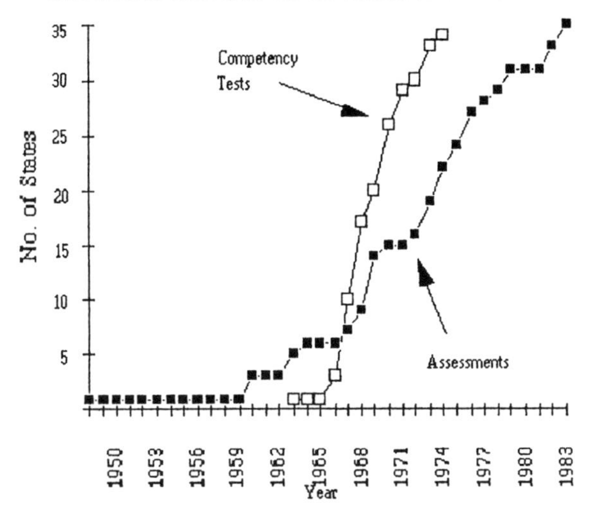

Figure 11.1 Numbers of states authorizing minimum competency testing and assessment programs
Source: Haney, Madaus & Lyons, 1993; OTA, 1987

involved in other types of entrance and certification testing. The principal source of revenue of NCS is test scoring, while Scantron markets test scoring machines to schools. Figure 11.2 shows the revenue trends for these four firms over these periods. For these four companies the period of the 1980s, after about 1982 or 1983, appears to have been one of unprecedented growth in sales for the testing industry.†

As indicated in Figure 11.2, the total revenues of ETS show a dramatic increase from $35 million in 1970 to $310 million in 1991, representing a compound annual growth rate of about 11%. Revenues of NCS have increased even more sharply over the last decade, zooming from $35 million in 1980 to $284 million in 1990—equivalent to a compound annual increase rate of 23%.‡ ACT has shown more modest revenues over the last 20 years, increasing from $8 million in 1972 to $70 million in 1991. This is equivalent to an annual rate of increase of about 12%, approximately equivalent to ETS's rate of growth in total revenues over roughly the same period. Between 1980 and 1988 Scantron revenues increased from $4 million to $35 million, representing an annual rate of growth of around 30%, considerably exceeding the growth rates of the larger firms.

---

† This trend is true also for those test publishers reporting sales figures to the Association of American Publishers (AAP).

‡ While detailed information on the breakdown of NCS total revenues is not available it appears that the vast majority of NCS revenues—of the order of 80–90%—come from scanning services, test building and test sales.

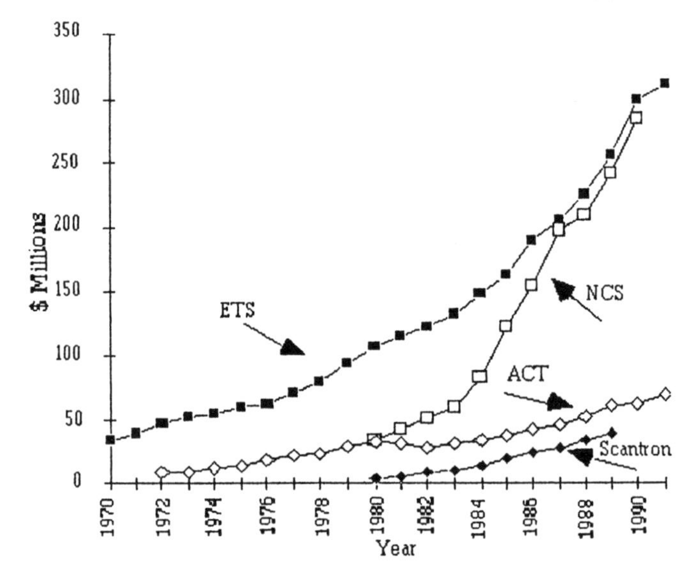

Figure 11.2 Revenues of four testing companies
Source: Haney, Madaus & Lyons, 1993; Company annual reports

Literature on educational testing as an indicator of growth

Haney developed another indirect way of documenting the increased attention to testing in the educational realm (Haney and Madaus, 1989). He charted the number of citations under the rubric "testing" (as indicated by number of column inches) from 1930 through 1985 in the Education Index. For comparative purposes, and because he argued that curriculum issues should be a central focus of schooling, the numbers of citations under "curriculum" were also charted.* These data are shown in Figure 11.3.

Figure 11.3 shows that the average annual number of column inches devoted to citations concerning curriculum has increased only modestly over the last 62 years—from 50 to 100 inches per year in the 1930s and 1940s to 100 to 150 in

* Figure 11.3 was constructed by measuring the number of column inches devoted to lines concerning testing and curriculum in every volume of the Index from 1932 through 1994. Over these volumes there were some changes in the index rubrics concerning testing and curriculum. These are the primary rubrics used by Haney, and which we used to update the figure:

Testing	*Curriculum*
tests and scales	curriculum
testing programmes	curriculum making
tests of general educational development	curriculum satisfaction
	curriculum selection
testing equipment	curriculum development (replaced curriculum making in vol. 11)
	curriculum studies

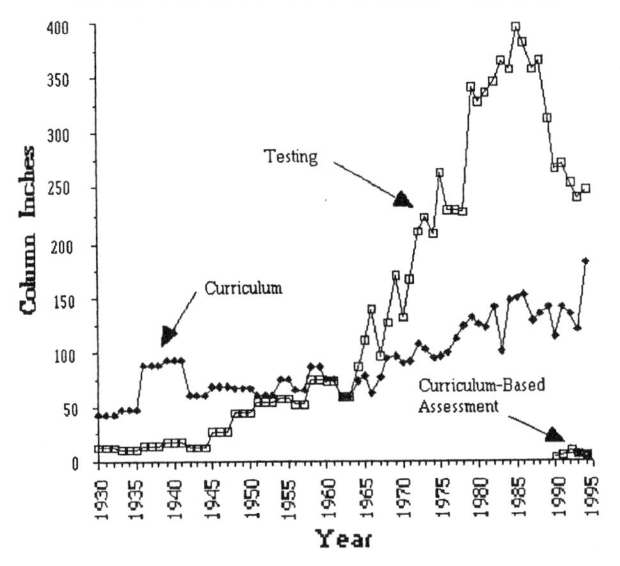

Figure 11.3 Education index listings under testing and curriculum
Source: *Education Index*, 1930–1994

recent years. In contrast, column inches devoted to tests and scales have increased greatly, from only 10 to 30 in the 1930s and 1940s to well over 300 in the 1980s. The past few years have seen a decline in the number of citations regarding testing; the new rubric "performance-based assessment", however, was added to the Education Index in 1992 to reflect prevailing testing terminology, and those citations are not included here. We do include data for "curriculum-based assessment", another new rubric that was implemented in 1990. While these indices are admittedly crude, the data certainly point to the fact that the prominence of testing, as represented in the education literature, has grown dramatically, particularly since the mid-1960s.

When the data shown in Figure 11.3 are considered along with the data on the growth in state testing programmes (Figure 11.1) and in testing revenue (Figure 11.2), the conclusion seems inescapable that educational testing in the United States has expanded vastly over the last 20 years, in terms of both volume and societal importance.

REASONS FOR THE GROWTH IN EDUCATIONAL TESTING

Why has the testing market-place grown so steadily? There is no single answer to the question. Instead, we consider the following four social forces that have combined to create the extended market for testing:

1. Recurring public dissatisfaction with the quality of education, and efforts to reform education.
2. An array of legislation, at both federal and state levels, promoting or explicitly mandating standardised testing programmes.
3. A broad shift in attention from a focus on inputs or resources devoted to education toward outputs or results.
4. The increased bureaucratisation of schooling and society.

While we consider these categories separately they are intimately related one to the other. For example, public dissatisfaction with the quality of education has produced legislation which in turn has contributed to increased bureaucratisation.

Dissatisfaction and reform

Since mid-century there have been at least five major cycles of educational reform, all of which have contributed greatly to the growth in testing, namely: (1) the Sputnik episode in the 1950s, (2) the Civil Rights and compensatory education movements in the 1960s, (3) the Scholastic Aptitude Test (SAT) score decline in the 1970s, (4) the education reform efforts that emerged in the 1980s, (5) the national education reform proposals of the early 1990s.

The launching in 1957 of the Soviet satellite Sputnik sparked widespread national concern over the competitiveness of the United States with the former Soviet Union in the space race and more generally in science and technology. One result of this concern was the passage in 1958 of the National Defense Education Act (NDEA), which contributed directly to a growth in testing (described below).

In the 1960s, the Civil Rights movement, and related efforts to improve education for the disadvantaged, also brought increased attention to testing and calls for educational reform in two quite different ways. On one hand test results of economically disadvantaged students were cited to bolster proposals for protecting the civil rights of minorities and increasing the educational opportunities of disadvantaged children. On the other hand, as civil rights legislation was passed, test results became the touchstone by which many of its mandates (for instance the success of compensatory education such as Head Start, Follow Through and Title I) came to be judged.

During the 1970s the diagnosis—that a variety of ills beset the schools—was common. And among the commonly prescribed remedies was to increase testing. Hence, as shown in Figure 11.1, the 1970s and 1980s saw marked increases in state assessment programmes. Perhaps the most prominent example of the connection between testing and concerns over quality of education was the furore over national declines in SAT scores (used for college admission). From 1963 to 1977, the national average SAT verbal score declined by nearly 50 points, and the mathematics average by about 30 points. This unprecedented decline sparked immense interest and debate over the quality of education high school students were receiving.*

In the early 1980s, low test scores were frequently cited to bolster calls for reform of the nation's schools.† The most widely publicised and influential of the education reform reports of the 1980s was *A Nation at Risk* (National Commission on Excellence in Education, 1983). The National Commission on Excellence in Education (1983, p. 5) ominously warned that "the educational foundations of our society are presently being eroded by a rising tide of mediocrity that threatens our very future as a Nation and a people". The indicators of this "mediocrity" were results from the National Assessment of Educational Progress (NAEP), the SAT score decline, studies of functional literacy, data from the International Association for the Evaluation of Educational Achievement (IEA) and test scores from the Department of Defense. Among the recommendations of the Commission (1983, p. 28) was the following:

> "Standardised tests of achievement (not to be confused with aptitude tests) should be administered at major transition points from one level of schooling to another and particularly from high [secondary] school to college [third level] or work. The purposes of these tests would be to: (a) certify the student's credentials; (b) identify the need for remedial intervention; and (c) identify the opportunity for advanced or accelerated work. The tests should be administered as part of a nation-wide (but not Federal) system of State and local standardised tests."

In the view of the Commission, test results were not only a worthy source of evidence in diagnosing the ills of the nation's schools, but also were part of the prescription for remedying those ills—a trend that is present in current reform proposals.

The 1990s saw another round of educational reform proposals, the most prominent of which grew out of an "education summit" where President George Bush and the National Governors Association (NGA)—headed at that time by then Governor Bill Clinton—announced six National Education Goals which were later expanded to the following eight to be achieved by the year 2000:

1. All children will start school ready to learn.
2. The high school graduation rate [completion of 12 years of schooling] will increase to at least 90%.

---

* A special panel convened to look into the score decline concluded that the decline in average national SAT scores had two causes. First, until about 1970 the fall-off in scores was most likely due mainly to "compositional changes" in the population of students taking the SAT. More students were taking the SAT, and more were of the type that tended to earn lower scores. Second, after 1970 the continuing decline in SAT scores was most likely due to factors unrelated to the demographics of the test-taking population. While the panel named six interrelated factors as probable contributors in the SAT score decline, the chief suspects were the deteriorating quality of schools and schooling (Wirtz, 1977, pp. 46–48).

† See Murphy (1990) for analyses of the broader reform movement of the 1980s, and Stedman and Smith (1983) for an analysis of early 1980s reform reports.

3. American students will leave grades 4, 8, 12 having demonstrated competency in challenging subject matter including English, mathematics, science, foreign languages, civics and government, economics, arts, history, and geography; and every school in America will ensure that all students learn to use their minds well, so that they may be prepared for responsible citizenship, further learning and productive employment in our modern economy.
4. The nation's teaching force will have access to programmes for the continued improvement of their professional skills and the opportunity to acquire the knowledge and skills needed to instruct and prepare all American students for the next century.
5. United States students will be first in the world in science and mathematics achievement.
6. Every adult American will be literate and will possess the knowledge and skills necessary to compete in a global economy and exercise the rights and responsibilities of citizenship.
7. Every school in America will be free of drugs and violence and will offer a disciplined environment conducive to learning.
8. Every school will promote partnerships that will increase parental involvement and participation in promoting the social, emotional, and academic growth of children. (Public Law No. 103-227, 1994).

Space does not permit a detailed discussion of these eight goals except to note two things. First, the goals were incorporated in 1994 federal legislation entitled the Goals 2000: Educate America Act (Public Law 103-227) and, therefore, are currently the law of the land. Second, the Goals 2000: Educate America Act calls for the establishment of the National Education Standards and Improvement Council (NESIC) which will certify "voluntary" national content and performance standards, as well as certify state tests geared to national curriculum standards covering "core subjects" of English, mathematics, science, foreign languages, civics and government, economics, history, and geography. These new tests—sometimes called authentic assessments to distinguish them from traditional multiple choice tests—are intended to be used at the national level to help check progress toward the National Education Goals for the Year 2000. What is clear is that the Goals 2000 Act represents the culmination of a trend apparent in many earlier educational reform proposals, namely the explicit use of tests as instruments of educational reform. (What is not clear at the time of writing in early 1995 is the ultimate fate of Goals 2000 as the Republican-dominated Congress revisits educational issues.)

Legislation

A more direct influence on the growth of testing has been legislation at both federal and state levels mandating various standardised testing programmes. We cannot summarise state-level legislation that promotes testing here. Nevertheless, we would be remiss not to note that, as we saw above, the numbers of state-level testing programmes increased sharply after the mid-1960s, and most of the state testing programmes were mandated either by state legislation, or by state boards of education under their mandate in state constitutions or legislation. Here we briefly mention five major pieces of federal legislation contributing directly and indirectly to the upsurge in the testing market-place over the last 30 years.

(1) The National Defense Education Act of 1958

As noted above, in response to the Russian launching of Sputnik, Congress passed the National Defense Education Act of 1958 (NDEA). The NDEA was an important landmark for the testing market-place in that for the first time the federal treasury became a source of funding for testing at both state and local levels. (Many of the testing provisions in the NDEA were eventually incorporated into the Elementary and Secondary Education Act of 1965, discussed below.) It is surely no coincidence that shortly after enactment of NDEA, revenues from test sales began to climb.

(2) The 1964 Civil Rights Act

The Civil Rights Act of 1964 mandated the Equality of Educational Opportunity survey (Coleman *et al.*, 1966).* This survey, discussed below, commonly called the Coleman Report after its chief author, helped to move the attention of educational policy-makers away from equal educational opportunity defined in terms of school resources towards a focus on educational outcomes as measured by tests.

(3) The National Assessment of Educational Progress (NAEP)

The National Assessment of Educational Progress (NAEP) formally began in 1968.† NAEP has not contributed much to the total volume of tests administered annually in the United States. However, it is an important development in the history of testing in the United States for three reasons. First, NAEP was the first occasion on which the federal government directly funded, on a continuing basis, the gathering of nationally representative achievement test data. Secondly, NAEP data were used to support calls for educational reform in the 1980s and 1990s described above, which in turn usually called for new testing programmes.

Finally, despite a very limited impact on educational policy-making in its first two decades, it now appears that by allowing state-by-state comparisons of results, performance on NAEP exercises may focus even more national attention

---

* The 1964 Civil Rights Act was landmark legislation in many regards, not the least of which was its impact on employment testing which we will not discuss here.

† See Greenbaum *et al.* (1977) for an account of the early history of NAEP. Though begun with private foundation support, since the late 1960s NAEP has been funded almost entirely by the federal government.

on test scores in the future (Haney and Madaus, 1986). Simple rank-orderings of states—a league table as it were—based on 1990 results were widely reported in the national press. This despite warnings from US Department of Education officials about the meaningfulness of such comparisons (Stancavage *et al.*, 1992). It seems likely that current developments in NAEP are likely to continue to contribute to the prominence of testing in American society.

(4) The Elementary and Secondary Education Act (ESEA) of 1965

The Elementary and Secondary Education Act of 1965 and its amendments up through the Improving America's Schools Act of 1994 passed by the 103rd Congress represent a watershed in the history of federal government involvement in education. The main idea behind ESEA was that for the first time the federal government would provide financial assistance to local educational agencies serving high concentrations of low-income families.

As noted above, since attention had been drawn to achievement disparities in terms of standardised-test performance by advocates of federal funding for schools serving economically disadvantaged children, it is hardly surprising that legislators ended up mandating that efforts to eradicate educational disparities should be evaluated using the same standard (see Kennedy, 1965). Thus, test results became the yardstick used to measure results of ESEA programmes and other interventions such as the Head Start Programme and Follow Through. Since almost every school district in the nation qualified for ESEA "Title I" monies, evaluations of Title I became a major impetus to use nationally normed achievement tests during the 1970s.

(5) The Education for All Handicapped Children Act of 1975 (Public Law 94-142)

Another major piece of federal legislation that contributed substantially to the demand for more testing was The Education for All Handicapped Children Act of 1975 (Public Law 94-142) and subsequent revisions. Public Law 94-142 and its accompanying regulations promoted tests and other evaluation materials to determine individual placement, to assess specific areas of need, and to evaluate the effectiveness of individual educational plans (IEPs) mandated under the law for special needs children.*

The focus on outcomes of schooling

The pattern in recent education reform reports—using test results to diagnose what is wrong with education and then prescribing more or new kinds of testing

* The testing of special needs students in order for local educational agencies to meet the requirements of Public Law 94-142 seems to have provided additional impetus to the market for tests. Between 1978 (the first year in which the law was fully implemented) and 1984, the percentage of students nationally receiving special education under federal law rose from 9 to 11% (National Center for Education Statistics, 1987).

as a cure—and the trend apparent in increasing legislative attention to testing, both illustrate a fundamental shift over the last 20 years in how people regard educational quality. For most of this century people have viewed school quality not so much in terms of test scores, but in terms of a wide range of facilities, resources and conditions associated with schools. Thus, the perceived quality of a school depended on such matters as how well it was funded, the quality of its physical plant, the characteristics of its teachers and the demographic characteristic of its students.

The release of the Coleman Report (Coleman *et al.*, 1966) helped to prompt a dramatic shift in the way people judged school quality. One of its most startling findings was that school facilities for minority and majority children were not greatly dissimilar. Even more surprisingly, the study found that whatever differences in facilities and school resources existed they had little if any discernible relationship with student achievement. This conclusion—which was popularly and inaccurately summarised as "schools don't make a difference"—aroused considerable consternation among advocates of education in general and advocates of equal educational opportunity in particular (see Madaus *et al.*, 1980, for a discussion of the Coleman Report and its aftermath). For our purposes it is important only to note that this clearly contributed to a shift in national attention toward the outcomes of schooling. This shift in focus away from school resources toward school outputs clearly contributed to the prominence of standardised tests in the US since the 1960s, and consequently has contributed to the growth in the testing market-place.

To provide evidence of the extent to which outcomes have become a prominent concern of educational policy-makers, we cite just one recent example from President Bush's America 2000 proposal (US Department of Education, 1991) that paved the way for the current Goals 2000 legislation (Goals 2000, 1994). In America 2000, there was a proposal for new American Achievement Tests as part of a "15-point accountability package" through which it was argued "parents, schools and communities can all be encouraged to measure results, compare results and insist on change when the results aren't good enough" (p. 11). One prominent advocate of America 2000, Chester Finn, in his book *We Must Take Charge: Our Schools and Our Future*, argues even more strongly that holding schools accountable for outcomes "is the only kind of accountability worth having" (Finn, 1991, p. 149).

Bureaucratisation of education

A phenomenon contributing to the upsurge in standardised testing over the last several decades is the growing bureaucratisation of American education in this century (see Wise, 1979 and Hall, 1977 for a detailed discussion). In this century the control of American schooling has become increasingly centralised; large numbers of local school districts have consolidated to form fewer but larger school systems; state education agencies in most states have increased their influence over local education authorities; and since the 1960s and the passage of ESEA, the federal government has played an increasingly prominent role in education.

Externally mandated tests or something like them are an absolute necessity for centrally and hierarchically organised schools. Foucault (1979) also argued that exams act as a mechanism to keep students—and now we might add, given the way exams are used, teachers as well—under surveillance and thereby give those who control the tests power over students and teachers. Standardised tests provide a means for categorising people, educational institutions and problems according to abstract, impersonal and generalisable rules. They expedite formal and impersonal administrative procedures. For instance, in many jurisdictions, if someone certified to administer a recognised intelligence test gives the test to a child using prescribed procedures, the child may be classified on the basis of the resultant score as being of normal intelligence or educably or trainably retarded, and then placed in an "appropriate" instructional group.

Tests surely provide one of the commonest of antidotes to selection of personnel on the basis of political connections, family background or social class. This universalistic feature of tests provides society with a mechanism for the allocation of opportunities on the basis of "objective" qualifications or "merit". Scores on standardised tests in modern bureaucratic society represent not just measurements but signs of merit. In education, as well as in employment, standardised tests very frequently provide the means by which personnel may be selected not on the basis of patronage, social class, gender or race, but instead on the basis of "objective" qualifications—or at least on the basis of "objective" test scores. People who believe in this universalistic/meritocratic view of tests would do well to read *The Rise of the Meritocracy* (Young, 1958).

Estimates of the costs associated with educational testing

As documented above, the level of standardised testing in elementary and secondary schools has grown enormously over the past three decades, partly because of the belief that standardised testing is so inexpensive (often direct outlays are less than $0.80 per student per test hour*). However, such reasoning is misleading. The total value of resources devoted to educational testing is much larger than the direct, observable costs of developing, purchasing, and scoring standardised tests.

There are significant indirect resource costs that should be included in any attempt to evaluate the social investment in local and state testing programmes. Haney *et al.* (1993) estimate that the total costs of standardised testing for state and local programmes are 4 to 65 times greater than their direct costs. In addition

---

* We detail our estimation procedure below. Please recall that Haney *et al.* (1993) defined a "test" as any separately scorable portion of a test battery. We discuss costs in the context of testing hours, i.e. the total level of resources devoted to one hour of test administration.

Table 11.3 Estimate of the direct costs of state and local testing programmes per student per test hour

State testing programmes			
Low estimate		Weighted average cost per student	$3.01
	Divided by	Number of tests/student	3
	Divided by	Number of tests/hour	1.33
	Gives	Direct cost/student/test hour	$0.75
High estimate		Adjust for 1 test/hour	
	Gives	Direct cost/student/test hour	$1.00
District testing programmes			
Low estimate		Average cost of one test booklet plus scoring services plus answer sheet per test hour	$0.77
High estimate		Average per student cost for a large urban school district	$5.19
	Divided by	Number of hours tested	3
	Gives	Direct cost/student/test hour	$1.73

Source: Haney *et al*., 1993; Office of Technology Assessment (1992).

to the obvious direct costs of testing, such as test purchase and scoring and reporting services, two types of indirect costs are associated with educational testing. Administrative or transaction costs are related only indirectly to the achievement of a policy goal, but are nonetheless still paid in cash from public or private funds. The second, and by far larger indirect cost, is the opportunity cost of the time that teachers, administrators and students devote to standardised testing.

Table 11.3 shows Haney *et al.*'s (1993) estimates of the direct costs of state and local testing programmes.* Table 11.4 shows their estimated indirect costs of testing in the United States.

Comparisons of Tables 11.3 and 11.4 show that the indirect costs of state testing programmes are 3.7 to 65 times larger than direct costs, while indirect costs of local testing programmes are 3.6 to 38 times larger than their direct costs. Further, teacher and student time comprise the largest portion of total costs, ranging from about 66 to 94% for both state and local programmes.

Table 11.5 presents Haney *et al.*'s (1993) estimate of the nation's total investment in state and district testing programmes in the late 1980s.

Table 11.5 shows that the nation devotes between $86 million and $4.7 billion worth of resources annually to state assessment programmes, and between $225 million and $18 billion to district testing programmes (in constant 1988 dollars).

* The low and high estimates of direct state testing programme costs differ on the basis of assumptions regarding the length of testing time; the low estimate assumes that one test lasts 45 minutes (implying 1.33 tests per hour), while the high estimate is based on one test per hour. The low estimate for district programmes comes from Haney *et al.* (1993) estimates of the price for four major test batteries; the high district estimate is based on an independent report of the average cost per student of a testing programme in a large urban school district.

Table 11.4 Estimate of the indirect costs of testing programmes per student per test hour*

TRANSACTION COST			
Low estimate		Total annual per student expenditure†	$5091
	Adjusted by	10.2% devoted to building operation and maintenance‡	$519.28
	Divided by	Number of hours/school year§	1260
	Gives	Transaction cost/test hour/student	$0.41
High estimate			
	Adjust for	Five hours in-class preparation time per hour tested¶	$2.50
STUDENT TIME			
Low estimate		Weighted average student shadow wage across all grades	$1.01
High estimate		Student shadow wage	$9.01
	Adjust for	Five hours in-class prep time per hour tested¶	
	Gives	Student cost/test hour	$54.06
TEACHER TIME			
Low estimate		Average teacher hourly wage¶	$22.25
	Divided by	Student/teacher ratio†	17.30
	Gives	Teacher cost/student/test hour	$1.29
High estimate			
	Adjust for	Five hours in-class prep time per hour tested	
	Gives	Teacher cost/student/test hour	$7.74
ADMINISTRATOR TIME			
Low estimate		Average administrators/school, times 50% involved in testing programmes	0.77
	Divided by	Student/administrator ratio†	317.43
	Times	Administrator hours/test hour	0.50
	Gives	Administrator hours/student/test hour	0.001
	Times	Average administrator hourly wage‖	$24.05
	Gives	Administrator cost/student/test hour	$0.03
High estimate			
	Adjust for	80% of administrators involved in testing and 1 hour of admin time per test hour	
	Gives	Administrator cost/student/test hour	$0.09

* This table summarizes data presented across several different tables in *The Fractured Marketplace for Standardized Testing*. All data are for 1988 and refer to public schools only. Indirect costs of testing primarily arise from valuing student, teacher and administrator time. For a more complete explanation of the derivation of these figures, see Haney *et al.* (1993, Chapter 4).
† Snyder (1991).
‡ The 10.2% figure is the average percentage of total expenditure devoted to building operation and maintenance over the period 1970–79 nationally in public schools (Snyder, 1991).
§ For teachers and students, 180 days times 7 hours per day is 1260 hours per school year. For administrators, we assume 200 days times 8 hours per day or 1600 working hours per year.
¶ US Department of Education, National Center for Education Statistics (1991).
‖ US Bureau of Labor Statistics, *Occupational Outlook Handbook* (1988).

Table 11.5 Total national cost estimates for state and district testing programmes (millions of 1988 dollars)

State assessment programmes		Low	High
	Number of tests	33.00	71.50
Divided by	Tests/hour	1.33	1.00
Gives	Total test hours	24.75	71.50
Times	Cost/test hour	$3.49	$65.39
Gives	Total cost of state testing	$86.38	$4675.39
District testing programmes		**Low**	**High**
	Number of tests	85.62	271.63
Divided by	Tests/hour	1.33	1.00
Gives	Total test hours	64.22	271.63
Times	Cost/test hour	$3.51	$66.12
Gives	Total cost of district testing	$225.40	$17 960.18

Thus the total investment in state and district testing is estimated to be between $311 million and $22.7 billion.*

Total national expenditure on elementary and secondary education in 1987–88 was about $169.7 billion (National Center for Education Statistics, 1991). Haney *et al.*'s testing cost estimates range from 0.18% to 13% of that figure. Their high estimate indicates that the equivalent of more than 10% of our national investment in elementary and secondary education is being devoted to the costs of testing. As a comparison, federal government revenues going to elementary and secondary education in 1987–88 were $10.7 billion (National Center for Education Statistics, 1991, p. 47). The high estimate, therefore, indicates that the cost of testing in elementary and secondary schools is equivalent to more than double the total federal expenditure for elementary and secondary education. Thus, testing is not simply a cheap activity for monitoring educational outcomes. Whatever the informational and motivational benefits of testing, testing influences the outcomes of education through its use of valuable and scarce resources such as student and teacher time.

CURRENT DEVELOPMENTS

The Goals 2000: Educate America Act reform bill, passed in 1994 by the 103rd Congress, has mandated research and development addressing "such issues as developing, identifying, or evaluating new educational assessments, including

* There is obviously a huge range in these low and high estimates. We should reiterate that the main cause is the costs of student and teacher time devoted to test preparation that are included in the high estimate. Note, however, that even in the low estimates, which include just testing time and administrative test preparation and value student time very conservatively, the value of student and teacher time still constitute the bulk of the real costs of testing.

performance-based and portfolio assessments which demonstrate skill and a command of knowledge" (Public Law No. 103-227, Title IX, sec. 931(d)(2)). More importantly, as noted above, Goals 2000 also creates a National Education Standards and Improvement Council (NESIC) drawn from the ranks of state governors, members of Congress, higher education representatives, business leaders and organised labour representatives for membership. NESIC is charged with reviewing and certifying "voluntary" national student content, performance and opportunity-to-learn standards, as well as ensuring that state assessments meet content standards as well as accepted professional and technical standards. The importance of NESIC must be viewed in light of the Improving America's Schools Act of 1994 Chapter 1 provisions, which mandate that assessments used in Chapter 1 be approved by NESIC. This process raises the stakes associated with test use, and calls into question the claim that the national standards are indeed voluntary.

Since the November 1994 Congressional election, there have been proposals from the new Republican majority in the House of Representatives to abolish NESIC, disparagingly characterised as a "federal school board", despite the earlier Republican and Democratic consensus on the need for national "high" or "world class standards". As an alternative to federal control of standards, some Republicans propose giving individual states more power to determine educational policy. Experience suggests, however, that when states attempt to use alternative assessment for accountability purposes, problems arise. Outcome- or performance-based education is currently under attack in Pennsylvania, Kentucky, Virginia, Alabama and California by conservative parent groups concerned that the movement incorporates values antithetical to their beliefs (Asimov, 1994b; Madaus, 1994). In fact, a 10th-grade performance-based state-wide test known as the California Learning Assessment System (CLAS) was discontinued, partly as a result of controversy surrounding literature selected as prompts for English test items. Conservative Christian critics also decried instructions asking students to express their feelings about stories they read during the test, arguing that it is not the role of the school to test children on their feelings (Asimov, 1994a). Whatever changes are made to Goals 2000—and it is clear that changes will be made—the use of tests as instruments of educational reform, we believe, will not be diminished. (At least one presidential candidate, former Secretary of Education Lamar Alexander, has denounced Goals 2000 as "one of the worst pieces of legislation in memory . . . [because of the] way it tried to wrest responsibility for what goes on in our nation's schools from the parents, teachers and local school officials" (Alexander, 1995, p. A14)). The struggle will be over who controls the tests and the actual content of the test exercises.

The promotion of outcomes-based accountability in a country that highly values individual achievement may arise from a vision of examinations as meritocratic—an "objective" way to ensure that educational benefits are distributed on the basis of merit. It is clear, however, that public dissatisfaction with education (including ascription of the country's perceived economic ills to problems with the educational system), federal and state legislation, an intensifying focus on educational outcomes rather than resources devoted to education, and an increasingly bureaucratised educational system, all serve to promulgate more testing in this country. On a broader scale, publicity surrounding US performance on international assessments in mathematics and science* has not only enhanced interest in testing, but also has prompted one of the national goals—that US students will be first in the world in science and mathematics achievement. Educational testing in the United States, it appears, will continue to be a growth industry, and current reform efforts promise even further development and growth. It will be important to chart the introduction of new assessment techniques over the next few years. The experience of some states in using assessment as a policy tool indicates that problems with efficiency, cost, standardisation, reliability and validity are emerging, and one state programme is close to returning to norm-referenced multiple choice tests. Whatever the modality or description—standardised testing or performance assessment—it will remain a key policy tool in American education.

ACKNOWLEDGEMENTS

Work on this paper was supported by the Ford Foundation (Grant 910-1205-1). The authors are soley responsible for the content.

* Although the performance of several individual US states exceeds that of other countries (Beaton and Gonzalez, 1993), as a whole the US has performed relatively less well than its industrial competitors on international comparisons of achievement.

CHAPTER 12

The Integrity of Public Examinations in Developing Countries

Vincent Greaney[1] and Thomas Kellaghan[2]

[1] *World Bank, Washington, DC, USA*
[2] *St Patrick's College, Dublin, Republic of Ireland*

INTRODUCTION

Public examinations play an important role in many educational systems throughout the world. They have a long history in their country of origin, China, and developed rapidly in the second half of the nineteenth century in many countries, including Germany, France, Great Britain, and Japan. The examination models that were developed in Europe subsequently became an important component of educational systems in many developing countries. The examinations were important for at least two reasons: they served the function of controlling the disparate elements of educational systems and, of more immediate relevance in the lives of students, performance on them was used to select individuals for scarce educational and vocational positions (Kellaghan, 1992; Madaus and Kellaghan, 1991). While examinations still play a key role in the distribution of educational benefits in many industrialised countries, their significance is much greater in developing countries, if for no other reason than that alternative opportunities for advancement are more limited.

The concepts of equity and meritocracy are closely tied to the institution of examinations. In Imperial China, a prime consideration was that the "all-important business of government must not be left to the accidents of either birth or wealth" (Hu, 1984, p. 7). This idea was maintained when examinations were introduced to European and other countries. Following the French Revolution, the principle of equality implied that all citizens would be in a position to compete for qualifications, employment, and wealth, and that advancement would be based on demonstrated achievement in school and in examinations rather than on the

Assessment: Problems, Developments and Statistical Issues.
Edited by H. Goldstein and T. Lewis.

social position of one's family (Eckstein and Noah, 1993). Similar motivation is evident in Britain and in India where examinations were introduced to replace practices of patronage and of buying positions in the civil service and armed forces (Montgomery, 1965). In Japan also, in the latter half of the nineteenth century, a competitive examination system was introduced to open educational opportunities to the whole (male) population (Amano, 1990).

It would probably be incorrect, however, to conclude that the promotion of equity or equality was the only or even main driving force in the establishment of examinations. The reforms of Frederick the Great and of Napoleon, for example, seem to have been motivated largely by a desire to build a modern unitary state, staffed on the basis of talent (Eckstein and Noah, 1993), while Gladstone's determination to effect change was influenced not a little by the obvious administrative shortcomings exhibited by the army and civil service during the Crimean War (Montgomery, 1965). It might also be incorrect to assume that examinations, whatever the motivation in their introduction, did in fact play a major role in the promotion of equity in the educational system or in the distribution of life chances. There has, in fact, been very little by way of formal investigation of the equity of examinations in terms of the traditional categories of gender, home background circumstances, or location of residence (see Greaney and Kellaghan, 1995). In so far as social mobility was a feature of life over the past century and a half, it would seem to owe more to the provision of schools than to the institution of examinations, which probably did little more than reflect students' access to, and earlier success in, the educational system. Our concern here is largely with school-based examinations; similar issues in relation to university-based examinations are discussed in Chapter 8.

In this chapter, our focus is not on the association between examinations and their fairness for groups of individuals, however those groups might be defined (e.g. in terms of social background, gender, or ability—see Chapter 6). Rather, we will examine the extent to which procedures to standardise the conditions under which examinations are prepared, administered, and scored are observed or violated. Only if such procedures are successfully implemented can we be sure that the integrity, wholeness, or soundness of examinations is maintained so that no candidate is placed at an advantage or disadvantage relative to other candidates because of unfair practice and that, as a consequence, the marks or grades awarded candidates are directly related to the ability that is being measured rather than to irrelevant factors or uncontrolled conditions.

That uncontrolled conditions arise from time to time will not come as a surprise to anyone with any experience of examinations, testing, or indeed any competitive enterprise. In the case of public examinations, such conditions are clearly recognised in the regulations of examination boards which use a variety of terms to describe practices that would interfere with the integrity of the examinations. The terms include "misconduct" (University of London Examinations Board), "dishonest conduct" (University of Cambridge Local Examinations Syndicate), "cheating" (Oxford University Delegacy of Local Examinations), "unfair practice" (Joint Matriculation Board), and "irregularity", "misconduct", and "dishonesty" (Uganda National Examinations Board). It seems reasonable to subsume all these terms under the term "malpractice" which is the one we will use in this chapter to describe any prescribed action taken in connection with an examination or test that attempts to gain unfair advantage or, in some cases, to place a candidate at a disadvantage. The action might be taken by an examination candidate, a teacher, a supervisor, an official or employee of an examinations authority, or indeed by anyone with an interest in the performance of a candidate. A candidate might engage in malpractice in the hope of gaining entrance to a higher level of education, while a teacher or a member of a candidate's family might be tempted to collude in such malpractice to assist the candidate achieve his or her goal. In the process, teachers may enhance their own reputations. The range of individuals involved in attempts to infringe examination procedures goes well beyond a student's family or school personnel. In some countries, examination corruption has become a business and individuals and groups engage in malpractice for monetary gain. As we shall see, the range of behaviours that has been used to interfere with the workings of examinations is very considerable and runs from the crude to the sophisticated and ingenious.

Much of our consideration of malpractice, both its execution and attempts to detect and prevent it, will focus on administrative and technical matters, which are the day-to-day issues that preoccupy examination authorities in developing countries. Concern with such issues, however, should not lead to a neglect of the fact that malpractice is also, and perhaps primarily, an ethical issue. While ethical aspects of testing and examining have recently received some attention both in the United States (e.g. Haladyna *et al.*, 1991; Mehrens and Kaminski, 1989; Smith, 1991a, b) and in developing countries (e.g. Ahmad, 1994), much work remains to be done in the area.

In this chapter, following a description of our sources of evidence, we shall describe forms of malpractice that have been identified in examinations. We shall then review the limited evidence that is available on the frequency of malpractice. Following this, we shall consider reasons for malpractice and outline some of the procedures that have been developed to detect it. Finally, we shall consider ways to control or prevent its occurrence.

SOURCES OF EVIDENCE

The sources of evidence that are available on the topic of malpractice in examinations are not very extensive. By its very nature, malpractice is something that is not easily investigated. Besides, our focus is on developing countries in which empirical research on any aspect of education is very limited. We shall use as evidence the papers (published and unpublished) of examination authorities, newspaper reports and articles, and information provided by individuals with considerable experience of examinations in developing countries. We shall also draw on our own experience in working with examination systems, especially in Africa and Asia. Altogether, we cite evidence from about 20 developing countries. Information of a confidential nature will be reported in a general manner so as to avoid revealing the identity of individual candidates, examination boards, and in some cases, a country. We shall also draw on some literature from the industrialised world, mainly from North America, where most work on the topic has been carried out, since this provides some useful conceptualisations of issues as well as interesting contrasts with conditions in developing countries.

In considering evidence from studies in industrialised countries, it is important to bear in mind that several differences can be identified between the conditions in which those studies were carried out and conditions in developing countries. Of prime importance are the general educational and social contexts in which examining occurs. Since the consequences for students of performance on examinations in industrialised countries are not as great as in developing countries, the incentives for malpractice are less compelling. On the other hand, since security surrounding examinations is generally greater in developing countries, opportunities to engage in malpractice are more restricted, though the incentive to do so may be greater. Apart from context, studies of malpractice in industrialised and developing countries differ in a number of ways. Most studies carried out in the United States were based on the use of multiple-choice tests, which are amenable to a variety of statistical approaches in the detection of malpractice. While such tests are used in some examinations in developing countries, open-ended formats are more common. Further, most (though not all) studies of malpractice in the United States were carried out at the college or college-entry level. In developing countries, malpractice appears at a much earlier stage in the educational system since examinations are used to determine who will gain access to the inadequate number of places available in secondary schools and sometimes even in primary schools. Finally, much of the evidence on malpractice in industrialised countries was obtained by self-report in questionnaires, while the evidence from developing countries is for the most part based on observation by examination or government officials or by newspaper reporters. It would be easy to conjure up a series of biases that might affect both types of evidence. These differences in the nature of evidence available from industrialised and developing countries point to the need for caution in applying the findings of studies carried out in the former to the latter.

FORMS OF MALPRACTICE

Malpractice is probably as old as examinations. At any rate, several contemporary practices, including copying from other candidates, impersonation, substitution of scripts, and prearrangements with examiners through bribery, were anticipated in the examination system of Imperial China (Hu, 1984; Miyazaki, 1976). Today, forms of malpractice vary from country to country, and may have increased in recent years in some countries, perhaps because of a growth in the importance of public examinations coupled with a deterioration in administrative efficacy and ethical standards (see Ahmad, 1994). Every malpractice involves a deliberate act of wrong doing, contrary to official examination rules, and is designed to place a candidate or group of candidates at an unfair advantage or disadvantage.

What is regarded as malpractice will in some cases depend on the form of the examination or test. For example, it is obvious that to bring books into an open-book examination is not malpractice. Furthermore, whether or not examination or test papers are released to the general public after students have taken them has implications for what is and what is not regarded as malpractice in the preparation of students for the examination or test.

Malpractice may occur at any stage in the examination process from development of the examinations through preparation of students, the actual administration of the examinations, and marking, to the issuance and use of results (see Ongom, 1990; Pido, 1994).

EXAMINATION DEVELOPMENT

Leakage occurs when the content of any part of an examination is disclosed prior to candidates taking the examination. Offending personnel include staff members of examination authorities, including printers, proof-readers and messengers, personnel employed to develop the examination (setters) or to determine its suitability (moderators), and school administrators. Since individuals will pay large sums of money for advance information on examinations, the temptations put in the way of those involved in examination development can be great.

Leakage, while often suspected, is difficult to document. Among documented cases is one in Pakistan, in which paper setters ran their own tuition centres where candidates, on payment of substantial fees, were granted access to at least part of the examination papers (Commission for Evaluation of Examination System and Eradication of Malpractices, 1992). In another case, in which one examination board selected one 10-question paper at random from a pool of four such papers submitted by the same setter, each paper was found to contain a common set of five questions which had been sold to a publisher and could be purchased. In Bangladesh, the 500-item pool from which items are selected each year was also found to have been illegally obtained and sold.

PREPARATION OF STUDENTS FOR TESTING/EXAMINATIONS

Test preparation may or may not be regarded as malpractice, depending on conditions surrounding test administration. It is generally accepted that teaching test-taking skills and the content and skills known to be covered in a test is acceptable. However, in the United States, when tests are not released to schools after students have taken them, the provision of practice on a parallel form of a test is not regarded as ethical (Haladyna *et al.*, 1991; Mehrens and Kaminski,

1989). In the case of public examinations, both in industrialised and developing countries, where examination papers normally enter the public domain after an examination, practice on earlier "parallel" forms is not regarded as unethical. Indeed, the general expectation and practice is that such examination papers will be used to guide teaching and study and that students will be provided with the opportunity of taking them. It would seem that anything short of presenting students with the actual test items or questions of an examination in advance of the examination is regarded as acceptable in the preparation of students. This seems to be the case even if such preparation tends to increase test scores without increasing student mastery of the content domain being tested, thus corrupting the examination as a measure of student achievement (see Madaus and Kellaghan, 1992).

ADMINISTRATION OF EXAMINATIONS

Impersonation occurs when an individual who is not registered as a candidate takes the place of one that is registered. It is usually difficult for impersonation to take place without the knowledge of the school principal or the collusion of a supervisor. It frequently involves university students or teachers taking the test for monetary reward or as a favour for a girl friend or boy friend. It can be difficult to counteract, especially in some Muslim cultures where faces are veiled, or where female candidates are exempt from having their photographs taken for identification cards, or where photographs of females are not readily available to males. The Lahore Board identified three cases involving young female graduates who were forced by their employer to take an examination on behalf of his family members. The same board revealed that a student from a poor background, who had just failed to gain admission to college on the basis of the results of a public examination, had subsequently acted as substitute for a student from a wealthy family and, in return, had been promised an upgrade of his mark and monetary compensation. Upon discovery, it was found that the instigator of the scheme, an examination board employee, had charged a fee over 20 times that offered the impersonator (E. Ahmad, personal communication, 1995).

External assistance involves individuals who are not examination candidates giving unauthorised assistance to candidates. Invigilators are frequently involved. They may dictate answers, work answers on a blackboard, circulate sheets of worked-out answers during the course of the examination, or act as couriers of material into the examination centre. In India examination papers have been smuggled out of examination centres and answers hurriedly prepared and thrown back through open windows wrapped around stones (Moore, 1993). Some people have engaged trained crows to supply material from outside to examinees in Bangladesh (Bethell, 1989). In the Philippines, due to lax security, answers made known to outside helpers during the course of an examination were smuggled into the centre in student sandwiches. Pakistan provides examples of two blatant forms of external assistance. In one case in Karachi, the answers to a multiple-choice test were broadcast using the public address system of a nearby mosque. In the second, in Baluchistan, external helpers used ropes to reach the examination centre on the second floor to render assistance to candidates. Modern technology increases opportunities for providing external assistance during examinations. Pagers have been used in China, and cellular telephones, watches, and receivers linked to external transmitters disguised as hearing aids in Pakistan (Commission for Evaluation of Examination System and Eradication of Malpractice, 1992).

Much more sinister, however, is the development in the Punjab in Pakistan since the late 1980s of a "mafia" that pays large sums of money to obtain jobs as supervisors in certain centres in which they can provide answers to examination questions to candidates. This may involve complicity with, or intimidation of, examination officials, police, and other functionaries. "Mafia" activity is so widespread that an increase in pass rates in examinations has been attributed to it (Ahmad, 1994).

Smuggling of foreign materials relates to the introduction of unauthorised material (e.g. notebooks, "crib notes", charts, and answer booklets complete with answers) into an examination hall. Material is frequently smuggled in pants, shoes, hems, and bras, or information may be written on parts of the body. This form of common malpractice can be difficult to counteract especially where traditions do not permit searching of underclothes (Pido, 1994).

Copying refers to the reproduction of another candidate's work, with or without his or her permission. Lack of a proper hall, inadequate spacing between candidates, poor-quality furniture, and laxity of supervision all facilitate copying.

Collusion involves unauthorised passing of information between candidates during the course of an examination. Traditional strategies include exchanging notes or scripts or throwing balls of paper containing material that would be of assistance to other candidates.

Intimidation occurs when examination officials, including supervisors and markers of papers, are physically threatened by people seeking support for individual candidates. Intimidation, like other malpractices, is not new. Over 30 years ago, internal school assessment as part of the public examination system in India had to be abandoned because of "pressures and fear" (Directorate of Extension Programmes for Secondary Education, 1960, p. 100). In recent years, intimidation seems to have become more common and is now an unpleasant fact of life in several countries, including Nigeria, Pakistan, and Bangladesh. It can take many forms, including assaults on examination supervisory staff (Inter exams begin amid rowdy scene, 1993), destruction of a jeep by a mob unhappy with the presence of an examination board official in Baluchistan (S. R. K. Badini, personal communication, 1994), placement of knives or guns on desks by candidates to deter close observation by supervisory staff, assault on an examination board chairman by a group (which included a politician) over his failure to change the location of an examination centre, and the storming of a centre by the wife of a government minister accompanied by a party of armed men to enable assistance to be given to her son (Commission for Evaluation of Examination System and Eradication of Malpractice, 1992). Intimidation by candidates has also occurred when students, angered by unexpected questions, led a walkout and compelled other candidates to do likewise.

Community involvement in intimidation is becoming a matter of growing concern. In Nigeria, the Registrar of the National Business and Technical Examination Board recently reported in a public lecture that two invigilators of the Joint Matriculation Examination in Delta State were taken out of the examination centre and shot dead by members of the community when they refused to co-operate in malpractice by candidates. The head of the community had already had a meeting with his people in which it had been agreed that individuals who stood in the way of the success of their children would do so at the risk of their lives (*Daily Times*, 1995). Given the occurrence of such behaviour, it is not surprising that fear of reprisal from members of the community in Bangladesh has deterred examination supervisors from reporting incidences of malpractice. In a part of the Philippines which had been subjected to political and civil unrest, staff from the national examination body were advised that their safety could not be guaranteed if they conducted supervisory visits of examination centres; no such warning was given to other state agencies. Other forms of intimidation experienced in Pakistan include abuse of journalistic power to influence decisions made by examination boards and teacher union pressure to ensure that nominees are appointed as supervisors.

Substitution of scripts involves the replacement of answer sheets handed out during the course of an examination with ones written outside the centre before, during, or after the examination. It has been known (in Nigeria) for an individual outside the examination hall to write the examination (for a fee) and then, in collusion with the invigilator, to pass it to the client's candidate. In another case, in Pakistan, examination board officials recovered the stamped, numbered, and unused answer sheets of a small number of candidates and brought them to a local house where the candidates, armed with books and supported by helpers, completed the papers. The papers were returned by the officials to the original bundle of scripts before being sent for marking.

Improper assignment involves the deliberate placing of candidates in centres under the supervision of corrupt officials (Commission for Evaluation of Examination System and Eradication of Malpractice, 1992). In a case in Sindh, a husband and wife were assigned as supervisors to a centre in which their son was appearing as a candidate (Top BISE official removed from office, 1994).

Ghost centres are fictitious centres which are established by corrupt examination officials. In these unsupervised centres (e.g. houses), candidates can complete their papers without supervision and with the assistance of books and helpers (E. Ahmad, personal communication, 1995).

MARKING OF EXAMINATIONS

Marker malpractice may be initiated by a candidate or it may occur as a result of the initiative of an examination authority staff member. Candidates may take the initiative in contacting markers by providing their addresses and phone numbers on their answer sheets, presumably in the expectation that a marker may wish to make contact with them at a later stage for a financial payment. Markers, too, may identify particular candidates and offer to do deals with their parents (Commission for Evaluation of Examination System and Eradication of Malpractice, 1992). Corrupt officials may take the initiative by revealing the names of markers to parents of candidates, again opening the way for intimidation or bribery, or they may intervene directly in the marking process. For example, in one instance, in Pakistan, the father of a candidate was able to arrange matters so that he could mark his daughter's paper, while in another a controller of examinations was found to have assigned some papers to examiners of his choice who had not been approved by the relevant board of studies. Under this controller's influence, some candidates who obtained zero marks were declared successful (Naqvi, 1993). Senior officials have also been known to remove or substitute the papers of some candidates (Erfan *et al.*, 1995; SHC issues notice, 1995).

Marker malpractice is not always intended to benefit the candidate. In one African country served by the University of Cambridge Local Examinations Syndicate, a marker who bore a grudge against a particular family deliberately lowered the marks of a candidate from that family (J. Saddler, personal communication, 1994). In another case, after a practical examination had been completed, the external examiner threatened to fail all the candidates in a prestigious private school, if a family member was not granted admission to the school.

The prize for ingenuity in effortless malpractice must surely go to the university lecturer in Cambodia who advised prospective students that they would require his assistance to pass the entrance examination, which he was prepared to provide for a consideration of $500. In fact, he played no role in selection and was simply gambling on the probability that some students that paid him money would pass. He was quite happy to refund money paid by students who failed (Asian Development Bank and Queensland Education Consortium, 1994).

Issuing certificates and use of results

Awards and certificates may be forged or a candidate's official final ranking or diploma may be unjustly enhanced by an official of the examination authority (E. Ahmad, personal communication, 1995). Falsification of results sheets by examination board staff has also been recorded (Commission for Evaluation of Examination System and Eradication of Malpractice, 1992; Erfan *et al.*, 1995). In Bangladesh and the Philippines, highly professional-looking fake diplomas have been prepared by skilled printers.

FREQUENCY OF MALPRACTICE

Interpretation of data on the frequency of malpractice in tests or examinations must recognise that probably only a small percentage of those who cheat are caught, while those who admit to cheating in questionnaires might also represent an underestimate of those that were involved (Bowers, 1964; Davis *et al.*, 1993; Haney, 1993). In view of these and other considerations, it is difficult to obtain precise data on the frequency of malpractice, either in industrialised or developing countries.

Haney's (1993) review of the findings of a number of studies in the United States over a period of time indicates quite wide variation in the numbers admitting to cheating at various stages of their educational careers, a variation that may be attributable in part to how questions were posed. Haney's conclusion is that some 15 to 25% of students in the United States admit to having cheated on tests or examinations, mostly by copying, during their undergraduate careers. Very few students said that they cheated frequently and students were less likely to cheat on tests or examinations than on other educational tasks, such as preparing a paper or bibliography, for which the criteria of cheating are anyhow less clear. The only test-related activity for which a relatively high frequency of malpractice was reported (33% of students) was getting questions or answers from someone who had already taken the same examination.

Findings from a recent study in Britain of a limited sample of undergraduates broadly confirm many of the findings of American studies. An inverse relationship was found between the perceived frequency and seriousness of cheating behaviours. Behaviours that were rated as occurring infrequently and as being most serious tended to be examination-related. One in eight students admitted having taken unauthorised material into an examination (Franklyn-Stokes and Newstead, in press).

In developing countries, actual malpractice, as distinct from self-reporting, on the basis of anecdotal evidence, seems to be endemic in some places, especially India, Pakistan, and Bangladesh. According to one observer, in India, "schools where cheating is not reported make newspaper headlines" (Moore, 1993, p. A14); a teacher in one school noted that "During yesterday's exams, every student was cheating" (Moore, 1993, p. A20). In Baluchistan, a headteacher reported that in a typical examination session, 100% of students cheat; even those who initially seemed to have no intention of doing so would start when they saw what was going on around them.

Data on malpractice indicate, as one would expect, that detection is less frequent than its occurrence. In Punjab, Pakistan's most populous state, a total of 13,231 cases were charged with using "unfair means" in 1991 (Ahmad, 1993). The practice seemed most prevalent among private (external) and rural candidates. Possession of unauthorised material ("helping material") and the unauthorised activities of markers were the most common forms of malpractice (Commission for Evaluation of Examination System and Eradication of Malpractice, 1992).

The number of students in Pakistan charged with malpractice appears to be dropping, perhaps because of concerted attempts to deal with the problem (Erfan *et al.*, 1995); in the country's largest examination board, Lahore, the percentage charged dropped from 2.2% in 1992 to 0.5% in 1994 (E. Ahmad, personal communication, 1995).

A study of the Uganda Certificate of Education for the years 1987, 1988, and 1989 also indicates that relatively few candidates are detected engaging in malpractice. The mean percentage of candidates detected over the period was 0.74 (Ongom, 1990). The most frequent forms of malpractice were external aid (59%), smuggling (27%), and collusion (11%). Only two cases of leaking of examination papers were detected.

Although these figures for detection seem small in light of the anecdotal evidence on malpractice, they seem to be considerably larger than the 250 candidates detected in 1994 by the Northern Examinations and Assessment Board in Britain in its administration of 1.5 million papers (Leake, 1995).

REASONS FOR MALPRACTICE

The level of malpractice in public examinations should be viewed against a series of educational, social, economic, political, and legal realities. Perhaps the most obvious occasion for malpractice is when examination results are used for competitive purposes and can have a profound and immediate impact on the course of an examinee's life. This situation is not unusual, and descriptions of conditions in Imperial China in the past and in developing countries today exhibit many similarities. For example, examination success in China has been described as providing the sole avenue to upward mobility and as bringing with it incomparably rich rewards (Hu, 1984). Similarly, it has been pointed out that, in contemporary India, where poor people have only a few avenues to becoming members of the relatively small middle class, one such avenue is examination success and high-school graduation, which can lead to a job; failure, on the other hand, will lead to drop-out and the same menial job as the students' parents (Moore, 1993). The extent of the problem facing students can be appreciated when one considers that fewer than 10% of candidates are admitted to secondary school in Malawi and Tanzania (World Bank, 1990) and less than 5% of an age cohort gain admission to universities in the vast majority of low-income countries (World Bank, 1994b). Given these situations, it is hardly surprising that malpractice occurs. Even in countries where opportunities are not so limited, as for example in Korea, the high value accorded university education as the only goal worth aspiring to means that individuals will do whatever it takes to get through the gates of top universities (Well-schooled in cheating, 1993).

The benefits that accrue from good examination performance are not limited to students. In China, a teacher's reputation is associated with student success on the national university entrance examination (Lewin and Lu, 1991), while it has been observed in Lesotho that good results increase a school's drawing power (Kellaghan and Greaney, 1992). In Kenya and Sudan, results of national public examinations are published annually to provide the public with an opportunity to compare schools. Any of these situations might tempt some individuals to engage in malpractice. Even in industrialised countries, school officials have been accused of manipulating the examination process to improve the school's position in published rankings (Clare, 1994).

These considerations might lead one to conclude that "the higher the stakes, the greater the likelihood of cheating and unethical practices" (Gipps, 1994, p. 153). However, this may not be so. In fact, findings in the United States (Bowers, 1964) and in Britain (Franklyn-Stokes and Newstead, in press) indicate that the relationship between the perceived seriousness of malpractice and its frequency is inverse. For example, students are more likely to cheat on less important tests (course quizzes) than on more important ones (midterm or final exams). Further, in developing countries, even when the stakes attached to performance seem uniformly high, there would appear to be considerable differences in the extent of malpractice from country to country. While there can be no doubt that high stakes are an important consideration in attempting to explain malpractice, the available evidence suggests that we have to look for the operation of other factors also.

Personal factors would seem relevant in this context. Hartshorne and May (1928) in their well-known studies of deceit assumed that, given favourable conditions, some individuals have a propensity to cheat. While this assumption may or may not be correct, we should not be surprised if we were to find that differences between individuals in such areas as self-concept, need for social approval, and ethical values were related to malpractice. We are not aware of any evidence relating to this issue in the case of public examinations; however, findings from the United States do point to a role for personal factors in the practice of cheating. Thus, dishonest behaviour has been found to be more common among some kinds of students than others. For example, a higher incidence was found among business students and students who were motivated to attend college more for reasons of material success than for reasons of curiosity (Haney, 1993). In so far as material success seems to loom large in the motivation of students in developing countries, we might expect some to engage in malpractice to achieve success.

We would expect situational factors to interact with personal ones to play a role in encouraging tendencies to engage in malpractice. In the United States, the role of stress and pressure for high grades has long been recognised as a correlate of malpractice (see Davis *et al.*, 1993). In developing countries, not only are high stakes attached to examinations, but quota systems often operate in the determination of pass rates. The quota is often determined simply by the number of places available at the next highest level of education, though additional quota systems to favour specific groups are in place in some countries, as, for example, Sri Lanka (Kariyawasam, 1993; UCLES, 1990), Fiji (Greaney and Kellaghan, 1995), and Malayasia (Tzannatos, 1991). If success is perceived to depend on the number of positions vacant in schools rather than on attaining a given level of performance, might this affect how students approach their studies? The answer would seem to be yes. It does not seem unreasonable to expect that the operation of quota systems, in which success is more a reflection of the availability of school places than of achievement, will lead some students to conclude that results are determined by fortuitous factors, to be little more than a lottery. In this situation, students may attribute success to forces beyond their control and as something that will not necessarily be achieved through hard work (see Hu, 1984). As a result, students may perceive expenditure of effort as dysfunctional (Weiner, 1994), with the result that they are poorly prepared for examinations and, when the time comes, see no alternative to resorting to malpractice.

It has been suggested that other situational factors may result in inadequate preparation of students for examinations and also lead to malpractice. Parents and students may perceive conditions for learning to be so inadequate that they have little option but to resort to unfair means. Moore (1993) concludes that students' cheating in examinations in India is due to the fact that they have not been taught throughout the year, because of chronic shortage and absences of teachers, lack of textbooks, and the poor physical conditions (clusters of open tents in 50 New Delhi schools) in which students have to work. Textbook shortage and inadequate teacher training, teacher absence from school, strikes, frequent school closures, and teaching that is of poor quality are features of other Asian and African countries. As far as teacher quality is concerned, it is worth noting that teachers in the Philippines (Philippines Congressional Commission on Education, 1993) and teacher trainees in Pakistan (Robinson, 1995) have been found to perform no better than their students on external achievement tests.

Inadequacies in the examinations themselves may also lead students to engage in malpractice. If examinations are designed to select only the highest scoring students, then it seems reasonable to assume that the examinations may not be sensitive to the achievements of less able students who may be able to deal with the situation only by resorting to malpractice. Ongom (1990) provides the example of a subject (history) in Uganda for which the syllabus was regarded as very wide and examination questions as being too general. As a result, teachers and students lost confidence in their ability to deal with the examination and, in desperation, took notes, notebooks, and textbooks into the examination centre.

Where governments or ministries of education do not enforce standards in other aspects of their work, it is difficult to expect that standards will be adopted or enforced in the public examination process. In Pakistan, appointments of senior officials to examination boards are based on political rather than professional criteria, a procedure that increases the likelihood of political interference in the running of the examination system. In developing countries, as elsewhere, politicians may be motivated by considerations of personal or sectional gain rather than the overall national welfare, and power may be exercised by ignoring or circumventing rules rather than by respecting them (Bhatty, 1995). In this context, the integrity of a national system can be undermined by an examination authority or minister for education that yields to political pressure to favour a certain candidate, that insists on special quotas of ministerial nominations to universities without reference to candidates' marks on the selection examination, or that fails to take action when evidence of malpractice is produced (as happened recently in one Asian country). In a more positive vein, a formal statement issued recently by the government of Uganda in support of the national examination board in a case of suspected malpractice sent a clear and necessary public signal that political pressure would not be allowed to influence the marking outcome.

A number of other factors, while they may not be regarded as reasons for malpractice, are likely to facilitate it. One such factor is the location of examination centres. In many developing countries, where centres are remote, material has to be dispatched up to one week in advance to ensure timely delivery, thus increasing opportunities of gaining access to examination papers. Furthermore, supervision of such centres tends to be infrequent. It is not surprising that a relatively high incidence of malpractice has been observed in centres in Baluchistan which are located over 600 miles from the capital in very remote areas.

Finally, the low level of salaries of teachers and of examination officials contributes to malpractice. In countries in which teachers' salaries are considered below the poverty level, bribes from parents of examination candidates may prove to be irresistible.

THE DETECTION OF MALPRACTICE

Supervisors, invigilators, and examination board officials provide much of the evidence of malpractice. Malpractice of various types has also come to light through complaints from candidates who were not involved and from others who became disillusioned by the failure of bribed examination personnel to deliver the expected level of support. Not all malpractice is detected, of course. It seems likely that some forms go unreported because they are generally regarded as acceptable behaviour. For example, paying an official in the expectation of gaining a special favour may be acceptable in some cultures, though it is officially regarded as a form of corruption by examination authorities. Cultural factors may also circumscribe the detection strategies that may be used. For example, it is difficult to obtain evidence that would support a charge of smuggling against female candidates in many African and Asian countries. Despite these difficulties, all examination authorities employ strategies to detect malpractice.

A number of procedures are used to ensure that only formally registered authentic candidates are allowed sit for an examination. In many countries, candidates are required to submit photographs at the time of registration which may be checked against the candidate at the actual examination. Thumb-prints are also used. While school principals and invigilators are often expected to ensure that only legitimate candidates are allowed to sit the examination, they do not always perform this duty in a satisfactory manner. One case is reported from Pakistan in which well over half the candidates who had completed some papers withdrew from an examination centre when informed that a team dispatched by the examination authority was about to check candidate identification documents (E. Ahmad, personal communication, 1995).

In Pakistan, external assistance provided by large bands of helpers around examination centres and within centres has been detected through unannounced visits by members of examination boards and by journalists.

Various approaches are adopted to find evidence of malpractice after an examination has been completed. When multiple-choice tests have been used, statistical indices are sometimes employed. Two approaches are possible. One is based on assumptions about the theoretical distribution of wrong answers; the other is empirical. For example, scores are examined to determine whether the number of wrong answers that are identical in two papers exceed the number that is likely by chance (see Angoff, 1974; Haney, 1993; Saupe, 1960). Although such probabilistic methods, which have been used mainly in the United States, may be appropriate in the study of patterns among populations, they are of limited value in determining malpractice in particular cases (Buss and Novick, 1980; Haney, 1993).

Most examinations in developing countries require more extensive written responses from students than is the case with multiple-choice items. When this is the case, the presence of neatly written material unrelated to questions on the examination paper in the middle of relevant hurriedly written answers may suggest that answers were prepared beforehand and smuggled into the examination. Other indicators of smuggling include answers written on paper other than paper that is officially provided and carefully written and well-illustrated answers with few spelling errors alongside poorly produced answers with many spelling and syntactical errors. Evidence of copying tends to be based on the presence of identical mistakes and peculiarities in scripts answered by candidates sitting close to each other. If collusion among candidates is suspected, reference is made to the seating arrangement of candidates recorded on the day of the examination. Sometimes, the detection of malpractice is fortuitous. In one case, an examination board was able to deny results to a school when markers discovered identical answers to the first 18 questions on a test that required short written answers. In response to the nineteenth item, one candidate had written, "Do this one yourself."

In Uganda, independent recorrecting of scripts revealed evidence of marker malpractice and of enhancement of marks by computer personnel. Marker bias against a particular candidate was identified by the addition of a second response to a series of multiple-choice items using a pencil of a different shade.

The National Education Testing Center in the Philippines has used prior information about level of school achievement to identify "statistically improbable results", especially in schools where marks appear to have risen sharply. In one case in which malpractice was investigated, five of the six highest scoring schools on a national examination had previously recorded relatively poor results (L. Tibigar, personal communication, 1995). In another case, the examination authority waged a successful court defence of its refusal to award examination diplomas by demonstrating a pattern of common errors among candidates in a school.

Malpractice does not cease with the issuing of certificates. Examination authorities have the responsibility of ensuring that their certificates are not interfered with or are not used by individuals who have not earned them. Strategies that are used to detect alteration of marks on an official diploma include using a code number on the diploma which contains information on the candidate's performance. For example, if a candidate was awarded 86 on a mathematics test, the code number might read 527648. The original mark can be calculated by subtracting each of the second and fifth digits from 10. Two examination bodies deliberately include errors in their diplomas in the hope that forgers might not notice them. In Ethiopia, when individuals came to the Ministry of Education some years after an examination to obtain a duplicate certificate for one they had claimed to have lost, their claims were checked by asking them to provide family information that had been recorded on the examination registration form.

WAYS TO CONTROL AND PREVENT MALPRACTICE

Control of examination malpractice is a never-ending battle. As officials succeed in counteracting one source, enterprising candidates will find new methods. To counteract this, severe penalties for malpractice have been used in the past, including the death penalty, confiscation of property, and exile for corrupt examination officials in seventeenth and nineteenth century China (Hu, 1984). In more recent times, lengthy prison sentences have been specified in Malawi and Kenya. In Korea, the illegal admission of students to university resulted in sacking of professors and, to punish the university, a reduction in the number of students admitted (Well-schooled in cheating, 1993).

Specific penalties are prescribed for different forms of malpractice in many countries on the assumption that the threat of being caught and punished will deter malpractice (see Tittle and Rowe, 1973). In contemporary China's selection examination for higher education, for example, 30 to 50% of marks are deducted for activities such as writing one's name on the test paper at a place other than the prescribed one, and for answering questions after the signal to stop has been given. Disqualification for one year is the penalty for whispering and copying, disqualification for two years for fabricating certificates or interfering with examination personnel, and for three years for impersonation. Malpractice by examination officials, bribery, or other serious offence is punishable either under the State Secrets Act or the Criminal Code by up to 10 years imprisonment (People's Republic of China National Education Commission, 1992).

Not all countries have adequate legal provision to deal with malpractice. In such cases, examination authorities may have little legal or political backing to support their efforts to punish students or adults. Even where the legal framework is in place, police and judicial authorities are frequently corrupt or for other reasons may show a marked reluctance to enforce sanctions. Where it is apparent that sanctions will not be enforced, the law serves little or no useful purpose. Politicians are unlikely to support punitive measures if the political authority is weak or if students are highly organised, as in Pakistan. Very often, politicians are part of the problem; a senior board official once confessed that since he owed his position to political support, he could not refuse any request from a politician. In Uganda, the attraction of corruption is lessened, and the status of members of the National Examination Board increased, by paying staff slightly higher salaries than personnel in comparable levels in the Ministry of Education.

Several procedures are adopted to prevent malpractice during the process of examination development. In Uganda, paper setters are required to set individual questions rather than an entire paper (Pido, 1994). In other cases, several sets of parallel papers may be produced from which a final one is selected. In Indonesia, five parallel versions of an examination are developed, and are assigned to schools on a random basis, thus reducing considerably the potential for leakage since schools cannot be aware which version will be assigned to them.

To guard against leakage, illiterate printers have been employed in Malawi, while many of the examination papers used in Kenya and Zimbabwe continue to be printed in England, despite the substantial cost. In Ethiopia, the typing of the Grade 6 and Grade 8 public examinations was entrusted to one typist. Prior to 1988, paper setters in China were held incommunicado for two months (Lewin and Lu, 1991). A similar procedure is adopted in Sri Lanka for printers and, in the Philippines, for printers and a handful of officials from the National Educational Testing Center who are responsible for supervising, proofing, and printing; these are not "released" until the examinations have been administered. Also in the Philippines, the contract with the printing agency includes a no-payment provision in the event of a leakage being traced to an employee of the agency. One examination board has insured itself for a seven-figure sum to cover the consequences of a leaked test. In general, tight security in the vicinity of the examination board, especially in the areas close to printing, packaging, mark tabulation, and computer processing, is regarded as essential.

Lack of roads and difficult terrain add to the difficulties of distributing examination material in many developing countries. To prevent leakage, papers are usually placed in sealed envelopes within metal or wooden boxes or sacks for transportation. Transit storage facilities include banks, police stations, and army barracks. In Kenya, government agencies, including the army and air force, may be called on to facilitate the delivery and collection of examination material. In the Philippines, a presidential decree requires all state agencies to support the administration of examinations. In an effort to ensure that the security of the process has not been breached, many examination authorities require packages of examination papers to be opened and answer sheets to be sealed in front of candidates.

To reduce the likelihood of collusion during an examination, supervisory staff are often drawn from schools other than the one in which the examination is being held. The authorities in the Philippines addressed the problem of copying by ensuring that candidates sitting close to each other were presented with different multiple-choice papers. While one row worked on the English test, the next responded to items in a different subject area.

Restriction of access to examination centres of unauthorised personnel, though sometimes difficult to achieve, helps limit the extent of external influences. Police are often involved in this, though, in some cases, they may contribute to the level of external assistance offered to candidates. In one region in Pakistan, examination supervisory officials have been given temporary magisterial powers, including the power of arrest. In many African and Asian countries, frequent random scouting missions to examination centres have been credited with reducing the level of assistance offered to candidates by supervisory staff. In one board's jurisdiction, the pass rate on the public examination dropped from 73.3% to 29.2% in one year following a concerted effort by senior educational administrators, board officials, and police to reduce levels of malpractice, which included a sharp increase in the number of visits to centres (Ahmad, 1993). In Cambodia also, a recent major campaign by the Ministry of Education to deal with malpractice resulted in a large drop (over 50%) in the pass rate in the school-leaving examination, a shortage of university entrants, and a large number of repeaters in the final school grade (Asian Development Bank and Queensland Education Consortium, 1994; Bray, 1995). A reform-minded Minister of Education persuaded cabinet colleagues to provide the equivalent of $US 1 million to pay supervisors and markers reasonable rates, ensured that no candidate was supervised by a person known to the student, provided police guards around examination centres, and safeguards for the secure writing and distribution of papers. Despite the collapse in the pass rate, this initiative received considerable popular support (V. McNamara, personal communication, 1995).

The advent of the fax machine poses an obvious security threat to examinations conducted on the same date in different parts of the world. One Pakistani board, charged with administering public examination to candidates in parts of Pakistan as well as to citizens working in other parts of Asia, Europe, and the Middle East, commences testing at 12.00 hours in Karachi, at 10.00 in Jiddah, and at 09.00 in Cairo to counteract the threat posed by the fax machine.

In both Bangladesh and Pakistan, the perennial problem of intimidation of markers by candidates and parents is being addressed by centralised marking and, even more effectively, by the use of substitute candidate numbers. In the latter case, original candidate identification numbers are replaced by other numbers and a record of the matching numbers is stored on a computer file. At the end of the marking process, the original and substitute numbers are matched.

Modern technology in the form of optical scanners and computers has the potential to contribute to a decrease in the incidence of malpractice. Optical scanners, when used to correct multiple-choice items, decrease the risk of improper interference at the marking or data-entry stages. Computers, because they reduce the number of people who are directly involved in the processing of results, also limit the opportunity for malpractice, especially at the results processing stage.

Jordanian examination officials capitalised on the computer to counteract efforts to have results altered by determining rigid cut-off points based on the results of the public examination to select entrants for various university departments and other highly valued career paths. When pressed by numerous external deputations to amend the cut-off points, officials were glad to be able to "blame" the computer for their inability to comply with the petitions (J. Socknat, personal communication, 1995).

As people become more familiar with computer technology, the Jordanian response is unlikely to carry weight. Further, the computer can be used to corrupt the examination process as well as to reduce the likelihood of malpractice. In Uganda, for example, a follow-up of the examination results for a school revealed that one candidate emerged with scores that were considerably higher than those awarded by the official markers. Investigation revealed that the scores had been altered at the final analysis stage by a computer operator (D. L. Ongom, personal communication, 1990).

Ultimately, success in preventing malpractice may depend more on the public's attitude towards it than on any technical procedure. Changing public perception will be difficult to achieve, however, especially when parents, teachers, and politicians feel that it is acceptable to use position and influence to achieve their goals. In an attempt to alter public attitudes, one examination board in Pakistan supported a national television drama series in which the consequences of malpractice in public examinations were portrayed (E. Ahmad, personal communication, 1995). Other strategies being considered by examination authorities in Pakistan include short radio messages and pronouncements from national sports heroes, religious figures, and senior political figures.

CONCLUSIONS

The high rewards attached to success in public examinations make malpractice at various stages of the examination process almost inevitable. In most countries, it should be said, it appears that the majority of candidates do not resort to the use of "unfair means". However, the evidence we have reviewed does suggest that examination authorities need to be ever vigilant in their efforts to ensure the integrity of the system that is used to certify that students have attained specified levels of achievement or to select students for the next highest level of the educational process. Failure to do so can have far-reaching negative consequences.

A candidate who knows that success on a public examination can be engineered by bribing or intimidating a marker, by purchasing a copy of the examination paper, or by using internal or external forms of assistance during the course of the examination, has little or no incentive to learn. It is particularly difficult for a student of high ability from a poor family to devote time and effort to study when he or she perceives that weak candidates from influential backgrounds routinely resort to unethical tactics to "succeed" in an examination. In such cases, far from being perceived as an instrument of equity, the examination system will serve to undermine students' belief that achievement and effort are rewarded.

Teachers are also affected by malpractice. There will be little incentive for them to work diligently to cover a broad curriculum if they have prior knowledge of the content of an examination paper or are aware that many of their students will resort to unfair means in the examination. In most countries considered in this chapter, teachers normally serve as examination invigilators. However, where malpractice at examination centres is perceived to be widespread, as in parts of Pakistan, many teachers feel that their reputations are being damaged by association with the examination and are refusing to participate in its administration.

At the national level, formal qualifications that are based on an examination system that lacks credibility will be viewed as being of little value. As a result, users of results, such as university admission bodies and especially employers, are being forced to create their own selection procedures. This can be an added cost and hence a disincentive for foreign investors wishing to establish industries or businesses in developing countries. Furthermore, individuals who have formal certificates or diplomas from third-level educational institutions that they did not merit pose a danger to society, especially in disciplines such as engineering and medicine.

Modern technology confronts examination officials with a set of new opportunities and problems (see Ahmad, 1994). On the positive side, computerisation, by limiting the extent of human contact with various aspects of the examination process, can help limit the possibility of malpractice. On the other hand, officials are faced with new challenges, including calculator-size instruments which have the capacity of holding vast quantities of information or can be used for transferring information from one candidate to another. Today, external assistance can be rendered by radio transmitters placed in pens, wrist-watches, and pagers, or linked with tiny hearing devices; by personal stereos loaded with prerecorded tapes; and programmable calculators packed with data.

The lack of good governance that has been a conspicuous feature of many developing countries is not unrelated to problems of malpractice in examinations. One often finds in the developing world weak institutional capacity, inconsistent and erratic policies, favouritism in the allocation of government posts, reward for political and personal loyalty rather than achievement, an undermining of professional standards and administrative duties, and a perception that the makers and enforcers of law are above it (Dia, 1993). In such circumstances, corruption and cheating can flourish, especially when examination officials are underpaid, when breaches of official rules are unlikely to be punished, and when opportunities for aiding candidates from wealthy or politically powerful families are abundant. Unfortunately, it would seem that in some countries, conditions favouring malpractice in examinations are becoming more common. Ahmad (1994, p. 1) has attributed the recent growth in malpractice in examination systems in Pakistan to "weak administrative machinery, changed social behaviour, politicisation of students and teachers, flow of weapons and political interference".

Efforts to deal with examination malpractice directly attributable to "official" interference should, as first steps, include reducing opportunities for bribery and intimidation, greater transparency in the administration process, support for a free press which should publicise evidence of wrongdoing, and a public education campaign to highlight the need for a credible and trustworthy public examination system. In many cases, legal reform may be necessary. Without the full conviction and political commitment of government, however, such reform will not be enough (World Bank, 1994a). Examination reform initiatives can also learn from experience in other countries. Co-operative regional efforts, such as those existing among African examination bodies, can be most useful in this regard.

No reform proposal can fail to take cognisance of the "social, religious, customary and historical factors in a society" (World Bank, 1994a, p. 27). This brings us beyond administrative and technical concerns and raises a consideration of cultural, and especially ethical, issues that may be relevant to malpractice. The limited available evidence suggests that actions regarded as cheating in one culture may be viewed quite differently in another. Within Anglo-American culture, for example, allowing another examination candidate to copy one's answers is likely to be considered cheating, while in Mexican and other cultures it might be regarded as a required form of altruism (Burton, 1976). A study of English as a Second Language (ESL) students noted that US students categorised the use of "crib notes", copying, and allowing others to copy as a more serious form of dishonesty than did Arabic and Spanish speakers (Kuehn *et al.*, 1990). In an Arab culture, cheating in certain cases may be explained in terms of a compelling desire to help a friend (Bagnole, 1977).

While to date, much of the literature on educational assessment has focused on technical aspects of measurement instruments, such as validity, reliability, and comparability, the evidence we have reviewed in this chapter suggests that malpractice represents as serious a threat to the credibility of the public examination systems of many developing countries as any technical concern. Furthermore, if malpractice is to be adequately addressed, this will involve more than dealing with the largely administrative and technical issues that are considered in this chapter. A satisfactory solution is not likely to be reached unless due consideration is given to traditional mores on the one hand and the needs of a modernising society on the other. Daunting as this task may be, failure to address the threats posed by malpractice in examinations, whatever their origin, is likely to have very serious negative consequences for the student selection and certification process, for teacher behavior, and for the quality of student learning. Given this situation, examination bodies and governments would seem to have little option but to intensify their efforts to minimise the level of malpractice in the conduct of their public examinations.

ACKNOWLEDGEMENTS

The authors wish to express their appreciation for assistance in the preparation of this paper to Ejaz Ahmad, George Bethell, Mark Bray, Parween Hasan, Vin McNamara, John Saddler, James Socknat, Yemi Suleiman, and Lucila Tibigar.

CHAPTER 13

Large-scale Assessment Programmes in Different Countries and International Comparisons

Leslie D. McLean

Ontario Institute for Studies in Education, Toronto, Canada

INTRODUCTION

As is abundantly clear by now, assessment is many things to many people, has a long history, and the form it takes in any jurisdiction is as much the result of political and economic forces and social values as educational concerns. Assessment by teachers in classrooms, especially in elementary schools, is a clinical, almost ethnographic process. Judgements are arrived at by observation, talk, written exercises, perhaps tests, often culminating in a written report on each pupil. An analogy can be made to qualitative research. Because each pupil's work is judged on its merits and the decision is whether it is up to standard (usually as defined by the teacher), this has also been referred to as the "standards" model (Taylor, 1994; Biggs, 1995). Distrust of this way of evaluating pupil achievement has led to many different attempts at assessment from outside the classroom, seeking to augment, even replace, subjective judgements by "measurements" believed to be objective—the "measurement" model (Taylor, 1994; Biggs, 1995). Teachers in secondary schools usually mix the models, but some move almost completely to the measurement model.

When the design and administration of assessment is moved outside the classroom, and especially when it is managed from outside the school for many schools, we call it large-scale assessment. Teachers often contribute some or all of the assessment tasks, but final assembly of the assessment instruments is done by other than classroom teachers so that none are advantaged. Responses from the pupils are entered into computers, analysed and reports prepared. Teachers,

Assessment: Problems, Developments and Statistical Issues.

officials, parents and community representatives may be asked to comment on the report, even to suggest standards of performance, but the calculation of summary scores on which judgements will be based is done by technical "experts". We will return to the matter of summary scores many times, but first an analogy for large-scale assessment.

Unlike the qualitative process of assessment by teachers in classrooms, large-scale assessment becomes a scientific endeavour which is, except for the nature of the measuring instruments, much like sub-atomic physics—quantum mechanics. Large-scale assessors seek to measure "achievement", a complex of mental characteristics (particles? waves?) that, like electrons, cannot be observed directly. As with atomic structure, the summary achievement scores in large-scale assessment are defined by the mathematical equations imbedded in the theory currently in use. We do not yet define achievement by systems of non-linear differential equations, but we are moving in that direction (see Chapter 4). Many fine qualities are claimed for these mathematical models, but one unfortunate result is that only the technical "experts" even begin to understand the numbers on which all key judgements are made, and with the most elaborate techniques, everyone has no choice but to believe that the theory and the calculations have produced numbers that reflect what pupils know and can do. Lest this analogy seem too far-fetched, consider the example of reading achievement.

Teachers read with and to the pupils, ask them questions about what they read and pupils write stories in their own words. Skilled clinician/teachers quickly learn where pupils are having problems, what they might most profitably read next and in this interactive process (involving speaking, reading, writing and listening) stimulate and foster more of this crucial meaning-making activity. Still, there are many stories of pupils fooling the teacher for a time and pupils reading books the teacher would predict are too advanced for them—no system is perfect. What mathematical model will be a proxy for reading achievement as understood in schools and by language scholars? The first model was simplicity itself: the number of correct answers to questions following short texts to be read by pupils, plus other equations that revealed where the pupil's score fell among scores of some representative sample of pupils at or near her or his score—norms. Only the technical "experts" could write down the equations, of course, but they prepared tables and charts that ordinary folk could use to decide how well pupils were reading (a popular chart in North America was the "grade equivalent", said to be the grade in school where most students were who had the same score as the pupil in question—or the same average as the group being assessed). There were many, many drawbacks to this model, as most readers will be aware, and it is not the purpose of this chapter to go into them. Most of us do not think of "number-correct" test scores and norm-referenced interpretations as a theory and a model, but they certainly are, and they are still around. For the most part, however, they have been supplanted by more complex models, such as the "item response models" described in Chapter 4.

In the Reading Literacy Study of the IEA (see "cross-national comparisons" below), from which we will draw many illustrations, the international experts decided to group their tasks (multiple-choice items) in three categories, Narrative, Expository and Documents, and to fit the simplest item response model separately to the items in each category, across all countries. This is the measurement model writ large.

To aid interpretation of such scores, we may have an "information function", and various "norm-like" interpretations have been put forward to give meaning to the scale values. As mentioned in Chapter 4, some use the term "item response theory" (IRT), but the item response models are not derived from any general theory of reading; they are content free, models that can be fitted to responses to any task that can be declared correct or incorrect or scored on a simple scale, provided you are willing to make the strong assumptions they require. Their popularity is due not to their theoretical basis, but to the properties of the scale scores (if the assumptions are satisfied), as will be discussed after giving some examples of assessment programmes. To large-scale assessors, item response models appear heaven-sent, though to critics the temptations to make what they see as questionable assumptions suggest a different source.

NATIONAL CASE STUDIES AND INTERNATIONAL COMPARISONS

We begin with national assessments conducted in two contrasting countries in South-east Asia, Indonesia and Singapore, continue with national assessments in the USA and Canada, and then consider Hungary, Israel and New Zealand. International comparative studies of education have been carried out since the early 1960s—essentially co-operative programmes which attempt to conduct national assessments in several countries at the same time, using the same instruments and the same design. The pace has picked up markedly since the early 1980s, and now one or two such "cross-national" studies are underway every year. The organisations that bring countries together for such ambitious efforts will be described and the studies listed, with references. Discussion of common issues comes next: sampling, non-response and partial response, special subpopulations, and background and context. For international comparisons we have to add the issues of language translation, funding and organisation and quality control. These last three can be issues in national assessments, of course, but they are not so salient as in the international studies.

Indonesia

For the second-largest country in the world (after Canada), the fifth largest in population, we look at the SLE, the school-leaving examinations, given nationally at the end of primary school (approximately Grade 6), junior secondary (Grade 9) and senior secondary (Grade 12). Pass/fail decisions are made after each examination, with the SLE score weighted equally with school marks. The score at the end of senior secondary is not used for post-secondary entrance. The examinations are thus important—at least moderately high stakes.

Consider the challenge facing the assessors (who are employees or appointees of the Ministry of Education and Culture). About 4 million students finish primary school, 2 million junior secondary and more than 1 million senior secondary each year. In principle, all should be tested (though see discussion below of special subpopulations). The schools are located on thousands and thousands of islands across a distance about the width of the USA or Australia. Tests cannot be given on exactly the same day in all places, so each test is prepared in several forms, which should be of equal difficulty and closely comparable content. (There are seven forms for each of six subjects in the current junior secondary SLE.) New tests must be prepared each year.

It is reasonable to ask (though it would appear no one has), "Can all this be done?" Tests have been given for some years, but the Indonesians became increasingly aware of the flaws (poor-quality tasks, non comparable tests, uncertain validity) and set out to correct them (Umar, 1994). They believe it is possible to have a fair and valid test every year for 2 million 12-year-olds, that it can all be done with the Rasch model, and they are well along with its implementation. Thousands of item writers have been trained and multiple choice items are piling up for each subject. The route to tests of equal difficulty, all testing the same content, is the Rasch-calibrated item bank: responses to items accepted by the professionals are collected in pilot tests and the Rasch model fitted—good items are the ones that fit, and vice versa. The essential requirements are: *unidimensionality* (a single trait, "ability", is measured by all items—see Chapter 4), and *trait dominance* (the probability of a correct response for every pupil depends only on the pupil's level of "ability" and not on the difficulty of the items or other factors). The first can be tested to some extent but the second cannot. A superb explanation of unidimensionality as used here is, "Unidimensionality, in this case, refers to the fact that the responses to the test items can be described with the use of a mathematical model *that assumes them to be generated or affected by one single dimension*, regardless of the psychological or cognitive dimensions actually involved in the construct" (Gonzalez and Beaton, 1994, p. 179, emphasis added).

Think of it: a single, unified measure of morality ("moral education" is one of the six subjects) valid enough to use in making pass/fail decisions for all those pupils from Timor to Java to the small islands in the West. Choosing the correct option on each item may indeed depend only on their level of morality, but the level of morality must be reasonably high since a crucial implication of the "trait dominance" assumption is that no one guesses at the answer. Ability to read the national language must also be ruled out (yet the Indonesian pupils scored near the bottom of the 27 countries participating in the Reading Literacy Study of nine-year-olds).

Why would reasonable people be willing to make such assumptions? Because if they do (and if their items can be fitted to the model), then comparable test forms can be assembled at will by drawing items at random from the item bank, and the bank can be constantly replenished. "Morality" can be expressed on any scale they wish; 0 to 1000 is a popular choice, with a mean of 500, suggesting great precision of measurement. An estimate of the standard error is easily produced by the computer program, and the estimates are always small. If I were the chief examiner, I would go for it, especially as the Rasch model comes recommended by professors at the University of Chicago, and they will train your graduate students to build and calibrate your item banks. Experience with mathematics and science would suggest that banks of calibrated items can be built, but banks for moral education and national language may be more difficult. In the Common Issues section below we will discuss one more powerful attraction of the model—the barriers to corruption and cheating it affords.

Singapore

National examinations are given to every pupil at the end of Grade 6 (prepared centrally, marked by teachers), at the end of Grade 10 (Cambridge Examination Board O levels), and at the end of Grade 12 for "academic students". The examination at the end of Grade 6 includes many multiple choice tasks, but also short answer, essay, speaking and listening tests. The tests for academic students are heavily multiple choice, with essays in language and some problems in mathematics and science. The Rasch procedure is used in item development and to calculate the final score.

United States National Assessment of Educational Progress (NAEP)

The NAEP was the first major monitoring programme to employ pupil sampling on a national scale. The 1960s brainchild of the late Ralph Tyler, its purpose was to assess the level of achievement by region, the closest anyone could get to the schools because of local autonomy and extreme sensitivity to any assessment from outside the classrooms at that time. Times change, however, and now state-by-state comparisons are being made for some states. Pupils were (and still are) sampled by age, rather than grade, age 13 being chosen as a time at or near the end of elementary school when virtually all students were still in school. Tasks were prepared and marked by the Education Commission of the States, located in Denver, Colorado, and percentage correct scores were reported by region, task by task. Little use was being made of the results, and after 20 years the Educational Testing Service (ETS) was given the contract as administering agency (see Chapter 11). ETS changed the definition of achievement to the now fashionable item response model (not Rasch: the model with difficulty and guessing parameters is used). Age sampling was retained. As this chapter is written, major changes are being made, so the form may be quite different by the time it appears. A succinct summary of the NAEP saga may be found in Lapointe (1986).

Canadian School Achievement Indicators Project (SAIP)

Canada has no national ministry of education, education being the sole and exclusive (and jealously guarded) responsibility of the 10 provinces and one of the two "Territories" (Yukon). As this is written, the federal government still administers education in most of the Northwest Territories. This dispersed responsibility explains why there was no national assessment of achievement until 1993, when the same world-wide pressures for accountability and lust for comparisons finally overcame provincial reluctance. So strong were the pressures that the provincial ministries of education (through their umbrella organisation, the Council of Ministers of Education (CMEC)) took the lead in organising SAIP—the only way it could have been accomplished. The way was opened by active Canadian participation in the OECD Educational Indicators Project (hence the name), which led to much expanded provincial co-operation in the definition of educational statistics other than achievement.

Given the history, it probably comes as no surprise that the first national assessment was a modest one—a sample survey of reading, writing and mathematics among 13- and 16-year-olds. With discussions just under way toward a national curriculum (unthinkable 10 years ago), a curriculum-based survey was not attempted. Such surveys are not completely out of the question, however, because Canada is participating in the Third International Mathematics and Science Study (TIMSS), a broad curriculum-linked study of the teaching and learning of mathematics and science at Grades 4, 8 and 12, organised by the International Association for the Evaluation of Educational Achievement (IEA), to be discussed below. The organisers of SAIP described the intent of the programme:

"The SAIP assessment is not a mass testing of students on a standardised test format. Because the CMEC assessment is criterion-referenced, emphasis is on gaining a portrait of the range of performances of groups of students of a certain age or gender, with overall program improvement as the objective. The aim is to help education systems know how they are doing and how to identify priorities" (CMEC, 1992, p. 1).

The organisers are able to claim "criterion reference" because assessment tasks were rated on a five-point scale by teachers. Just how education systems are to identify priorities has never been spelled out. More recently, "literacy" and "numeracy" surveys have been conducted.

Hungary

Until 1986, Hungary followed a traditional Central European path, with school-based evaluation until the mandatory matriculation examination at the end of secondary school. A majority of pupils had left school for jobs or vocational training by this time, so the examination was an academic one. Monitoring crept in by the international back door in the early 1980s, when a small team at the Centre for Educational Evaluation organised Hungarian participation in the IEA Second International Mathematics Study. A programme of monitoring every four years was started in 1986, but it did not continue until 1991.

When dominance by the Soviet Union collapsed in 1989, the existing structures were abandoned and a new research institute was organised. The Centre for Evaluation was moved intact, and the Director of that Centre is now the equivalent of a Deputy Minister of Education. Clearly, monitoring had proven its worth and the lessons learned were being applied in areas other than mathematics. In 1991, Hungary participated in the International Assessment of Educational Progress (IAEP II: mathematics and science), organised by ETS, and in the IEA Reading Literacy Study (9- and 13-year-olds), using these international studies as a base to conduct a national assessment of reading comprehension and mathematics. The national assessment was repeated in 1993, adding computer literacy and science. Hungary is currently active in TIMSS, with data collection to take place in 1995. Among all countries, Hungary has gone the furthest in integrating international studies with its national assessments.

Israel

The traditional Matriculation Examination (*begruth*), mainly essay tasks, is still required for successful completion of secondary school, but there is also a multiple choice examination very similar to the US Scholastic Aptitude Test, with verbal and numerical scores. A score is also given for English as a Second Language, and the test is sufficiently used and well regarded that schools have sprung up to prepare students—for a fee. Like the SAT, it is primarily seen as a predictor of success in post-secondary education. Schools give a final examination (the *gemer*), which is not sufficient for entry to university but can help entry to vocational schools and does ensure that all who complete secondary school get some sort of certificate. In short, Israel has no policy or plan for national assessment, but has participated in several international studies (SIMS, SISS, IAEP II) that were in effect national (sample survey) assessments. Regarding SIMS and SISS, see under Cross-national comparisons below.

New Zealand

For a small country without extraordinary economic resources, New Zealand has the most ambitious assessments of all. Since their participation (and leadership as co-international co-ordinators) in SIMS in the early 1980s, they have experienced dramatic change of government (huge cuts in public expenditure and increases in taxes) and widespread educational reforms that include a national curriculum and national assessment. They are going further to change the form of assessment than others, however, by requiring of the assessment contractors that the assessment tasks have "intrinsic motivation" (that is, they are interesting to the pupils) and that they embody new assessment strategies of interest to teachers. Attention to motivation is especially needed because the stakes are low—there is no threat to students, teachers or officials.

This all sounds interesting but not in the reform mould, and by itself it is not likely to satisfy those who demand school-by-school accountability (if not teacher-by-teacher) in the form of test scores. Perhaps for this reason an all-pupil assessment is planned for ages 5, 11 and 13. At age 5, the class teacher is to prepare a dossier for each pupil after six months of schooling and this dossier is to be reviewed by principals and parents. For ages 11 and 13, pools of tasks are to be prepared and made available to schools from a central computer. The Ministry of Education will then choose half the tasks for a national test and each school will choose the other half. All items are to be linked to the national curriculum. Regional trials were expected to begin in July and August 1995.

CROSS-NATIONAL COMPARISONS

The first large-scale international comparative study was organised and conducted in the early 1960s by a new organisation created for the purpose in 1959, the IEA, or to give its full title, the International Association for the Evaluation of Educational Achievement. The studies organised by IEA always include in-depth analysis of curricula, school organisation, teacher and student attitudes and background and socio-economic indicators. Since that first study (elementary and secondary mathematics, 13 countries), the IEA has sponsored the Six-subject Survey (science, reading, literature, civic education, and English and French as foreign languages, 8–19 countries, completed in 1973–74), the Second International Mathematics Study (SIMS: Grade 8 and Grade 12, 20 countries, conducted in 1981–82), the Classroom Environment Study, 10 countries, 1982), the Written Composition Study (13 countries, 1984), the Second International Science Study (SISS: 24 countries, 1984) and the Reading Literacy Study (31 countries, 1992). Ongoing studies are of early childhood care and education (the Pre-primary Study, 15 countries), Computers in Education (20 countries) and the Third International Mathematics and Science Study (TIMSS, more than 50 countries). The first 25 years of IEA studies are described in Postlethwaite (1987) and later work in the UNESCO journal, *Prospects* (1992). Goldstein (1995b) provides a critical evaluation of the work of the IEA including a commentary on its technical procedures. The IEA has completed the preliminary organisation of two new studies: The Language of Education Study (LES) and Civic Education.

Studies are approved and monitored by a six-member Standing Committee and given formal approval at the annual General Assembly, the thirty-sixth of which was held in the Latvian capital of Riga in September 1995. IEA has published an annotated bibliography of articles and books written about its studies containing over 700 entries (Degenhart, 1990).

In the late 1980s several countries approached the ETS for permission to use some of the NAEP tasks in their national assessments. When the number of countries passed three, someone suggested that they band together and carry out a common study. Several of the Canadian provinces joined in and the first International Assessment of Educational Progress (IAEP) was born, funded largely by a US government grant to ETS. Data were collected on mathematics and science achievement in 1988 from 13-year-olds in Canada (four provinces), Ireland, Korea, Spain, the UK and the USA, using NAEP multiple choice tasks agreed upon by representatives from all participating countries. Only a little background information was collected, and the final report (ETS/IAEP, 1989) set a new speed record by appearing the next year. The attractive and readable report was widely praised, though some technical faults were found. Planning for a second such study began immediately, and IAEP II surveyed mathematics and science achievement for 9- and 13-year-olds in 20 countries in 1991 (Lapointe *et al.*, 1992).

The Organisation for Economic Co-operation and Development (OECD)

The passion for comparisons of educational systems sweeping the world in the 1980s led the OECD in 1988 to launch an ambitious project to design and collect comparable statistics from all its members. Three categories of indicators were chosen: (1) Costs, resources and school processes; (2) Contexts of education; (3) Results of education. Student outcomes are part of the third category (along with system outcomes and labour market outcomes), the first version of which has R1: Performance in reading, R2: Performance in mathematics, R3: Performance in science, and R4: Gender differences in reading achievement (Bottani, 1994, p. 337). Results from IEA studies have been used initially for these "performance" indicators, especially the Reading Literacy Study, which included most OECD member countries. With more than 50 countries participating, the Third International Mathematics and Science Study is expected to yield those indicators in 1996 or 1997; in the meantime, the Second OECD Report will have only two achievement indicators. The high costs and long preparation times international studies require remain as significant obstacles to the regular production of educational achievement indicators.

COMMON ISSUES

Sampling

Sampling is an issue even when every pupil is assessed (no sampling of pupils or schools) because the tasks are inevitably a tiny sample from all possible tasks. When all pupils are tested, the sample is smallest, since the number of tasks is the maximum any one pupil can be asked to answer. The goals of individual testing and broad curriculum coverage assessment can perhaps both be met if everyone is willing to adopt the assumptions of item response models, because tests incorporating different tasks (with some overlap) can be "equated" using the model; the scores are treated as equivalent even though the tasks are not the same. Success at equating is not guaranteed, however, and the resulting tests might be too narrow or biased as a result of task selection in the equating process. If the stakes are even modestly high, the risk of controversy, even litigation, is high if examinees or their parents find out that the tests were not the same. This issue is important in large-scale assessment because interpretations are complicated unless the scores are stable over schools and regions (must be "generalisable", as the experts put it). Here is a dilemma:

1. The tasks are selected by subject experts to achieve reasonable coverage of the target curriculum (reading, mathematics to Grade 8, . . .);
2. Technical experts say that more or different tasks must be used, or some dropped, in order to meet technical standards (model fit, reliability, generalisability).

Do the assessment managers accept reduced or changed curriculum coverage in order to meet technical standards, or is coverage (huge factor in validity of achievement scores) more important than stability?

If scores are not stable over schools and regions, the multilevel models described in Chapter 10 may be used to estimate the extent and nature of the complexity, and many argue that this is the way to resolve the dilemma. Generalisability can always be achieved by making the tasks simple and/or averaging over many different tasks, but the resulting score may be meaningless. Assessments everywhere are moving in the direction of the New Zealanders, however, mainly by including more and more "performance tasks" (longer, more complex tasks requiring pupils to work out and construct responses), and IEA requires more tasks than any pupil can be expected to answer (Linn, 1994). In monitoring studies, broad curriculum coverage is attained by the use of multiple test booklets, each with a different set of tasks; pupils respond to only one or two of the booklets assigned at random. Provided classes have at least 10 pupils (20 is much better—and very common), class averages can be estimated for each task and aggregated to provide summary statistics based on random samples of student responses.

The issue is even more important for individual testing—certification examinations for example. "If it is expected that the results of an on-demand assessment consisting only of tasks requiring, for instance, a half hour or longer will be used to certify individual student achievement and substantial consequences will be attached to those certifications, then the generalisability results obtained to date are quite serious" (Linn, 1994, p. 10). We are only just coming to appreciate how very crucial the sampling of tasks is to the validity of our assessments, and to the stability and accuracy of the results. The certification examination in Indonesia is not likely to consist of performance tasks, not with one million students to be certified—large-scale assessment indeed; the issues in such assessments are different and will be discussed below.

When the purpose of assessment is monitoring the general levels of achievement (at the school or local authority level, for example, or the country), then a well-designed sample of schools and pupils is by far the most efficient method. NAEP, SAIP, New Zealand monitoring and all of the international studies are of this type. As brought out in Chapter 10, however, well-designed samples are several-stage cluster samples that are very tricky to design and analyse. Here is the brief description of the sampling procedure for the IEA Reading Literacy Study (Elley, 1992), where 27 countries are represented in Population A (average age of pupils in the sample was 9.7 years) and 31 in Population B (average age between 14 and 15). It illustrates complex grade-level sampling (in contrast to samples by age, for example) and gives a sense of the ways national samples must be designed.

> In each country, Population A consisted of the students in the grade level containing the most 9-year-olds, and Population B consisted of the students in the grade level containing the most 14-year-olds . . .
>
> To obtain comparable samples of students, multi-stage sampling was used in each country and schools or classes were typically drawn with a probability proportional to the size of the school or class. Where schools were drawn, an intact class was selected at random within each school, but in Population B some National Research Co-ordinators selected students at random from all classes in the grade level in the school.
>
> To overcome fluctuations in the execution of the sampling, weighting was used to adjust for any variations in the probability of selecting students. These sampling weights were used in all data analyses.
>
> In three of the smaller countries, all students in the relevant grade level were tested. (Elley, 1992)

Sample survey design requires special expertise, and hence a "Sampling Referee" is appointed to consult on and approve the samples for each country. Readers familiar with the field will recognise the name of Leslie Kish, the Sampling Referee for the IEA Pre-primary study and one of the most respected experts in the world. The importance of the process can be seen in the reference to "sampling weights", used in the calculation of final scores in order to make the results as comparable as possible from country to country. National samples also have to be "weighted" when subpopulations are over- or under-sampled.

Good co-operation from schools is essential if the designed sample is to be achieved, and in most countries the schools can decide whether or not to participate. Low participation (less than 80% of schools and students, but more than 70%) resulted in England's results being separated from the main analysis in IAEP II, for example, with the caveat, "interpret results with caution because of possible nonresponse bias" (Lapointe *et al.*, 1992, p. 18). Data from China, Portugal and Brazil were also separated because their samples were not comparable to the others.

When the data are in, cleaned, aggregated, weighted and (usually) scaled, there is still the step of estimating the standard errors of the summary measures to be used in the final reports. Little statistical expertise is needed to anticipate that the formulas for standard errors will be horrendously complex, if they can be written down at all, so they are usually calculated using an approximate technique, the "jacknife", a creation in the era of large, fast computers which involves resampling the data many times and carrying out parallel analyses for each sample from which the required standard errors, etc. can be obtained. The jacknife can be used at any level of aggregation, with sampling over tasks, for example, rather than over countries. The alternative, multilevel modelling approach, as described in Chapter 10 is now coming to be preferred since it is statistically more efficient and also provides important information about the variation among schools which the simple summary measures do not contain.

Non-response and partial response

Missing data due to non-cooperation have been mentioned under sampling, but data can also go missing when pupils respond to some tasks and not to others. Large-scale assessors attempt to design test booklets so that virtually all pupils have time to complete all tasks. This is essential when multiple forms are used and is always desirable. The purpose of monitoring is to sample the best performance from pupils, not to rush them through the tasks. When multiple choice tasks predominate, the usual procedure is to ask pupils to answer all tasks, even if they are not sure of the correct option, because guessing is always possible and all pupils answering all tasks brings us as close as possible to a level playing field.

However much planning and pilot testing is done, some pupils will not finish in the allotted time, and some will not answer some of the questions. Assessors must decide whether to ignore the non-responses and calculate averages based on only those tasks answered or to mark the non-responses as incorrect. Quite different outcomes can result from this choice, especially if item response models are being fitted.

The problem becomes acute if there are substantial differences in non-response patterns between or among subgroups (males and females, low and high scorers, ...) or, more likely, from one country to another. In the Reading Literacy Study the instructions to pupils were clear, that "they should answer every item (task), that they should work quickly, and that they had 40 minutes to complete the test" (Elley, 1992, p. 99). In spite of these instructions, substantial non-response was found from a few countries (where there was, perhaps, an unwillingness to guess) and a non-trivial number of incomplete answer sheets (suggesting that the pupils were not able to finish the test). Marking the missing responses as incorrect significantly lowered the overall averages in some countries, but this scoring was done in view of the instructions, the obvious compliance in most countries and the extensive pilot testing that suggested the time limits were reasonable. Here is how the author summed up the argument (Elley, 1992, p. 100):

"It is recognised that the style of response to formal tests may differ from one country to another. Thus it is possible that scores may have been higher in countries which had many students who did not complete the test. Their traditions may have encouraged them not to respond when they were unsure. For these countries, the scores in this booklet may be an underestimate. Under the circumstances there is no ready way of estimating their scores precisely."

So there it is—large-scale assessment often presents dilemmas—but choices have to be made.

Special subpopulations

Whether testing every pupil or monitoring, the large-scale assessor in every jurisdiction finds some pupils that cannot be assessed with the instruments prepared for the majority—pupils whose language is other than the national language (ONL) are nevertheless in school, and language-limited pupils (LLPs) some of whom used to be labelled "retarded" and were in special classes are now more often to be found in the "mainstream". In national assessments, schools often quietly excuse whichever pupils they judge excusable, but when accountability is the goal and stakes are high, some supervision and operational definitions are needed to quiet the temptation to exclude many who are neither ONL or LLP, but who are known to perform poorly on tests. Temptation can become practice when schools follow instructions to test everyone (for diagnostic purposes and to avoid stigma) and then find themselves criticised for their low average in comparison to others who were less scrupulous or who have a more advantaged population. Unfortunately, the publication of raw-score league tables without supplementary information is still common.

In international studies there has to be detailed negotiation about excluded populations, country by country, and some odd choices turn up. In the Reading Literacy Study, for example, pupils in French private schools were excluded—16% of pupils in Population A and 21% in Population B! The French could not be persuaded to include them (or could not persuade them to participate), and the debate whether the results from France should be set aside reached the IEA General Assembly. In the end the French results were included with the rest, but always with a footnote. In the schools of French Belgium, all students instructed in Flemish or German were excluded.

Sometimes *in*clusiveness fails to work. In countries such as Canada, where English *and* French are official languages, instruments have to be available in both languages. The province of Ontario participated in SIMS (Canada/Ontario), and instruments adapted to Canada from France, scrutinised and approved by Franco-Ontarian educators, were administered in schools where the language of instruction was French (about 10% of pupils attend these schools). After all the results were in, others made a case that the instruments in French were more difficult than those in English and the English/French comparisons had to be cut out of the national report at the last moment. The means were not recalculated for the international report, even though removal of the scores from the French schools would have marginally raised at least some of the scores.

Background and context

It is now widely accepted that valid inferences from test scores cannot be made without considerable knowledge of the *local* context for the scores, that is without some information on the nature of intake to the schools, the social and economic conditions in the area from which the pupils are drawn, and above all the pupils' "opportunity to learn" (OTL). Proficiency in the national language is influenced as much by contacts out of school as in school, but most school subjects are only learned in school. The mathematics and science in textbooks and on tests is not learned to any great extent from peers or family—or television. It is therefore crucial to know what has been taught—not what was *intended* to be taught, as in official curriculum guides for example, but what was *actually* taught, what has become known as the "implemented curriculum". Up to now, OTL has been estimated by asking teachers whether they taught the content or skills needed to respond correctly to the tasks, but a more ambitious scheme is being tried in TIMSS. In study after study, OTL has proved to be more highly correlated with achievement scores than any variable other than a prior achievement score.

Some national assessments still publish raw scores but addition of school characteristics and statistical adjustments for intake are beginning to be seen. Few go so far as Singapore, where in September 1995 the Grade 10 results were published for each school in relation to the Grade 6 scores in a form of "value added" analysis using multilevel techniques. Information was also given about the sports and music programmes of the schools, and, only in Singapore, the "obesity index". It was not obvious from the report in the *Straits Times* whether a high or low score was desired on the latter.

Fairness: comparability and barriers to corruption and cheating

In the Indonesian example, mention was made of the need for multiple forms of each test each year and new tests every year. Because the tests are used for selection and certification (weighted equally with school marks), they qualify as at least moderately "high stakes" tests. The credibility of the whole system is in doubt unless the different forms given as the "same test" are "comparable" (very similar content and difficulty). The administration of the tests and the collection and analysis of the test responses should be done in such a way that it is difficult or impossible for anyone to falsify the results. These requirements are especially challenging in Indonesia, but they are present to some extent in all national and international assessments.

If (a very large "if") all the assumptions can be satisfied, and if the item responses can be fitted to one of the item response models, then all the comparability requirements can be met: test scores for individuals (or countries, depending on the design) can be calculated on a single scale, whatever test form is used, and the scales can be equated across forms and from one year to the next. Moreover, after the hard work is done and the costs of preparing the bank of scaled items have been met, the random sampling of items and the complex scoring calculations make it difficult for anyone to tamper with the results. The item bank can be augmented and maintained at a moderate cost, compared to the cost of constructing new tests every year. As one celebrated psychometrician has put it, "The model is too good *not* to be true."

The problem with such discussions is that they beg the most important question of all, "Can there possibly be one numerical score that has the same useful meaning for all pupils in all parts of a country (local authority, province, ...)?" Agreement was achieved on three international reading literacy scales only after much debate; no one suggests that the scales be used for individual pupil assessment, but the country averages have been used to describe "effective" reading programmes by examining programmes in countries with high scores and those with low scores. Belief in some practices has been strengthened and doubt thrown on others.

Language translation

From the beginning, the language of work in the IEA has been English, even though its first headquarters was in Sweden. Survey instruments are prepared by international specialist committees in English, and when agreement is reached and the instruments are approved by the project Steering Committee, each instrument must be translated into the national language in each country by native speakers who know the subject being tested. In the early years, forward and back translation was urged, that is, from English "forward" to the national language, then independently from the national language "back" to English. The two English versions could then be compared and where they differed the translation reconsidered. Two translations were one too many for a number of countries, and IEA did not then have the resources to assist, so a single translation, carefully checked by bilingual educators, often had to suffice.

An international test of Reading Literacy presents the translation problem in its fullest complexity. The definition of reading literacy was, ". . . the ability to understand and use those written language forms required by society and/or valued by the individual" (Elley, 1992, p. 3), and here is how the assessors described their approach to translation:

> "Any translation process entails the possibility that the meaning is lost in translation, and an international test is no exception. To guard against this type of bias, test items were translated independently from the source language by native speakers *according to common guidelines, then compared and revised.* Every effort was made to ensure that the original sense of the text and items was maintained. Moreover, the items were examined after the pilot test and again after the final test *to ensure that they behaved similarly in different countries*" (Elley, 1992, p. 8, emphasis added).

There is no longer mention of back translation, and there are now "common guidelines" (not printed in the report). More reliance is being placed on expert judgement in the translation process, and technology is brought to bear after the pilot tests and even after the survey itself. Many more items (tasks) were given in the pilot tests than were required for the survey and the one-parameter (Rasch) model was fitted to the item responses separately for each of the three scales (narrative, expository and documents). Any poor-fitting items were dropped—about half of them, as it turned out. This is the way they ensured that the items "behaved similarly in different countries", one part of the rationale being that if the translation had changed the meaning of an item for one country then the item would not fit the common model. The logic is far from airtight, because models can fail to fit for many reasons, but if they do fit the case for common meaning across countries is strengthened. In the final analysis, a few items did not fit in one country but did in others; such items were not used to calculate mean achievement in that country, but the country score could still be estimated from the Rasch model. What the common meaning *is* has now to be re-examined, since the choice of tasks has now been influenced by technical, rather than substantive criteria (see also Goldstein, 1995b). An argument was made for Reading Literacy that the tests were broad samples of ability to understand what is written (less on ability to use the knowledge) and above all that the tests were comparable from country to country (Elley, 1992, pp. 4, 5, 8 and Appendix B).

Quality control

Whether national or international, large-scale surveys require the co-ordinated efforts of thousands of people in hundreds of schools. Any hope of comparability rests on the diligent administration of the instruments (as many as 100 different booklets, questionnaires and answer sheets in some international studies) in much the same way everywhere. The need for supervision of the sample *designs* has already been mentioned, but equally important is sample *execution*: preparation of the lists of schools (and ages of pupils for age samples) by region, community size and school size (or whatever strata are used) and competent application of computer programs. When many countries are involved the maintenance of high quality becomes a major task. The IEA relies on the national centres, supervised and guided by an International Co-ordination Centre (ICC) established for each study. For Reading Literacy, for example, the ICC was at the University of Hamburg. The National Foundation for Educational Research (NFER) is providing international co-ordination for the new Language of Education Study. The ETS co-ordinated the IAEP.

In all studies now, quality control representatives are sent to national centres several times, especially at the time of data collection. The ICC provides a computer program for the recording and checking of responses to the many forms and offers training in its use, this after painful experience in earlier studies when virtually unusable data sets arrived at the ICC from even highly developed countries. Huge efforts had to be made to retrace steps just at the busiest time for the ICC. Hard experience taught that investment in quality control throughout the project was much more effective (and efficient) than scrambles to clean up the data sets at the ICC. As studies get larger and the stakes get higher, good quality must be demonstrable. For a thorough discussion, see Schleicher (1994).

Funding and organisation

Although already touched on in passing, the matter of money and operational control deserve explicit mention. National assessments have often been narrow in coverage of curriculum and types of learning, high costs cited as the reason for poor coverage. As a voluntary association, the IEA depends most on its members to carry out its elaborate studies, but must raise very substantial sums from donors for overall co-ordination. The international costs of the Reading Literacy Study, for example, were US $615 000 over 42 months (Elley, 1992, p. v), a sum that does not include some of the generous infrastructure support provided by the University of Hamburg. Just as with earlier studies, the international co-ordinator and the IEA staff were constantly scrambling to raise the money needed to keep the study going. The enormous TIMSS is expected to cost 10 times that much, but financial support was obtained from the start from the governments of Canada (through the Ministry of Employment and Immigration) and the USA (through the National Centre for Educational Statistics, administered by the National Academy of Sciences).

Success in funding TIMSS may be bitter-sweet, however, because funding for the LES has not yet been found, in spite of widespread interest and universal admiration for the design of the study. Because of generous support from The Netherlands government for the IEA Secretariat in The Hague and dues of US $10 000 per year paid by most members (less from developing countries and $30 000 from the USA), IEA provided stopgap funding for Reading Literacy and travel money for national research co-ordinators and General Assembly representatives from less developed nations. Some start-up funds have been given to the ICC for LES. IEA members contribute in kind (and receive credit against dues) by organising and paying much of the cost of General Assembly and project meetings.

Up to now, the IEA (and for a short time, ETS) has decided which studies to organise, how to design them and what to report. TIMSS was proposed as TIMS (mathematics only), however, and science was added at the request of the US government in order to fit their plans for educational reform through the year 2000. If IEA studies are to continue to contribute to OECD indicators, then OECD is likely to want to help shape both the timing and design of studies. The IEA doubtless hopes that, as with the USA and TIMSS, requests for a share in decision-making will be accompanied by a substantial grant of funds. There are powerful arguments for leaving management and direction of studies with the independent IEA: (1) keeping the results virtually free of political influence and hence enhanced credibility of results (assuming adequate quality control); (2) the demonstrated capability of the IEA to bring together the foremost experts in the field from around the world to design and supervise their studies. Continuation, however, will depend on the achievement of stable financing.

AN ASSESSMENT OF ASSESSMENT

Comparative international studies have two important goals, according to the Chairman of IEA: to improve understanding of educational systems and to help understand the causes of observed differences in student performance. Improved understanding includes knowledge of achievement that contributes "to setting realistic standards for educational systems, as well as monitoring school quality" (Plomp, 1992, p. 278–9). National studies are far more likely to be seen as essential tools in educational programme evaluation, and they are often asked to evaluate the progress of individual pupils at the same time. These two purposes cannot be pursued well in the same survey, especially under the measurement model (Taylor, 1994). Large-scale assessments have proved a disappointment as tools for educational reform, leading to calls for new forms of assessment, especially performance assessment (Linn, 1994; Taylor, 1994). One frequent contributor to the literature writes, "The theory and practice of assessing learning are currently undergoing a paradigm shift", arguing that the measurement model is giving way to the standards model via a shift from quantitative to qualitative standards. The SOLO Taxonomy (Structure of the Observed Learning Outcome) is proposed as a replacement for quasi-continuous measures of traits, and applications to performance tasks is straightforward (Biggs and Collis, 1982; Biggs, 1995). The form of large-scale assessments is changing in the USA as a result of legislation requiring the use of performance tasks, both by states and the federal government, a strong emphasis on standards and requirements for the assessment of opportunity to learn (Linn, 1994). Performance-based assessment has a longer history in the health professions than in education in general, but there too the "formidable equating and security problems" have been noted (Swanson *et al.*, 1995, p. 10).

Calls are being made for performance assessments because our studies so far have not helped to set realistic standards. We cannot set realistic standards for teachers with unrealistic abstract traits, but paradoxically, the traits might be the best compromise. The Reading Literacy scales have been widely accepted as an achievement indicator: schools should therefore strive to attain higher scores on all three of them. The scales, however, tell teachers nothing they do not already know (include both narrative and expository texts and utilise a variety of documents). The only way to score well is to have an excellent reading programme! That is the extent of the backwash effect. (For a book-length exposition and spirited defence of the measurement model for monitoring and standard setting, see Tuijnman and Postlethwaite, 1994, esp. Chapters 9 and 10).

At the national or school level, links can be made to educational policies from general traits (such as the narrative, expository and documents scales). Hints as to the causes of differences in performance are sought through correlations, or what are essentially correlations: looking at how high-achieving countries differed from low-achieving countries, for example, in length of school year, access to books, amount of reading to pupils by the teacher and the like, with and without adjustment for social and economic differences. The Reading Literacy data have been analysed with admirable thoroughness and imagination, including the possible influence on achievement of more than a dozen different characteristics under the control of schools or authorities. Of 18 variables in Population A, only two had no correlation with achievement at all: class size and teacher with same class more than one year. Only the adjusted (partial) correlation was large enough for consideration for two variables: more formal tests and earlier age of entry to school. The discussion is clear and caveats appropriately numerous, and these analyses have been followed up in another book (Postlethwaite and Ross, 1992).

CHAPTER 14

Vocational Assessment

Alison Wolf

Institute of Education, London, England

INTRODUCTION

For most people, most of the time, assessments and qualifications are important to the extent that they help them do better in the labour market. Enjoyment of learning may play a part—but is not the uppermost consideration in the minds of French or English parents discussing with their teenage children which baccalaureate programme to follow or which GCSE (General Certificate of Secondary Education) options to choose. No more is it primarily the quality of the course design or teaching that persuades able students to pursue places at Tokyo University, Harvard or the Polytechnique.

To this degree, therefore, any form of educational qualification can be viewed as "vocational"—which the *Oxford English Dictionary* defines as to do with "employment, trade, or profession". Moreover, as more and more jobs draw on the general skills of literacy and numeracy, rather than on specific manual skills, the academic school curriculum becomes more of a direct, substantive preparation for work than was the case when large proportions of pupils were bound for jobs in agriculture or traditional mining and manufacturing industries. Nonetheless "vocational education" and "vocational assessment" retain, in ordinary parlance, the idea of something quite distinct from, and with lower status than, the "ordinary" educational curriculum.

This divide has its roots in historical developments which are common throughout the West. Until quite recently, the academic curriculum taught in *lycées*, gymnasia, grammar and independent public schools, lycea, etc. was the preserve of a small proportion of the population, the only route into a highly selective university system, and the precondition of entry into the traditional "professions" of the church, law, medicine, and the higher civil service. Its status was thus assured. The bulk of the population received a basic "elementary" education ending in most countries at 13 or 14, which concentrated on literacy and arithmetic skills, though with some practical elements added: needlework (for the girls) and woodwork (for the boys) thrown in. Such technical schools as

Assessment: Problems, Developments and Statistical Issues.
Edited by H. Goldstein and T. Lewis.

existed occupied a middle status, although most specifically technical and craft skills were taught outside the formal school system, through apprenticeship, "mechanics institutes" and the like (Sanderson, 1993). The following description of nineteenth- and early twentieth-century German education could have been applied equally well to any of the major countries of the West:

"general education and specialized training, that is, the acquisition of skills and knowledge for the sake of their direct application in an occupation, were supposed to be institutionally separated... [O]nly schools giving general education contributed to the cultivation of an individual's humanity, whereas those institutions which served to qualify their pupils for the practice of an occupation did not. The separation between general education and vocational training corresponded to a distinction between the social strata which partook of this education and this training." (Max Planck Institute, 1983, p. 239)

As secondary schooling lengthened and was offered to all, however, it became increasingly differentiated, with specifically "vocational" options increasingly available. The first comprehensive high schools were in the United States—also the first developed country to experience a large proportion of the population graduating from high school and entering college. However, the United States high school curriculum was never undifferentiated—on the contrary, three tracks were common from early on, with the vocational track incorporating "shop" or workshop classes. Many European countries—notably The Netherlands and France—developed specialist vocational schools which concentrated on craft skills, and which initially enjoyed considerable respect as a training ground for skilled artisans.

However, over the post-war period, general secondary education and post-compulsory study became increasingly the norm: a pattern which the United Kingdom adopted fairly late but which is now evident throughout the industrialised world. In the process, the specialised vocational tracks have become increasingly low status, rejected by parents and pupils. This is true even in countries with a tradition of well-respected vocational schools. For example, in The Netherlands the secondary level vocational schools (for age 14–16) which were the traditional choice of artisan families, are now "the last option . . . observers expect the collapse of junior vocational training." (Dronkers, 1993, p. 22). In Germany, while the apprenticeship system remains strong (see below), there has been a rapid fall in the numbers and proportion of students who, at secondary level, enter the *Hauptschulen.* These offer a very practically oriented curriculum, and no possibility of progress into higher education: and are overwhelmingly rejected by young people and their families in favour of a more general secondary education in the *Realschulen* or *Gymnasien.*

A result of these developments is that "vocational" has come to mean something which is essentially manual, and derives from a craft tradition, does not link into higher education, and is relatively low status. This obscures the important fact that, while there is a coherent concept of "vocational" education and assessment to be advanced, it is one which is specific to manual skills or lower-status occupations. The essential features and dilemmas of craft assessment are shared by the professions and the major contemporary issues in the field are of relevance across the occupational spectrum.

DEFINING "VOCATIONAL" ASSESSMENT

The distinguishing characteristic of the "vocational" derives from a key concept in assessment: that of *validity*. The conventional definition of an assessment's validity is the degree to which it actually measures what it is supposed to, and naturally everyone agrees that validity is a Tremendously Good Thing. Less obvious is how "validity" is actually to be defined and measured (see Chapter 6).

In the case of many academic subjects, the definition of valid assessments becomes, in practice, a circular process. It is the assessments which themselves define what is "supposed" to be measured—what the course is about. Debates over university finals papers in English or classics, or over the nature of the International Baccalaureate Mathematical Studies paper, are less about the validity of the *assessment* than the content of the course, and whether it could be improved. Given a set curriculum, a examination is valid to the degree that it samples from—tests—that curriculum. Validity of vocational assessments, by contrast, is not a circular concept and is not defined by the content of a course leading up to a test. It derives from something outside the process of teaching and assessment: and *an assessment may be defined as vocational to the extent that its validity has an external referent that derives from the world of employment.*

For example, if you want to judge the validity of a surgeon's final examinations, or of a driving test for heavy goods vehicles, you will look not at the syllabus which led to the examination or test (as you would with many school or university examinations) but at the behaviour and practice which follow after it. Does the assessment in fact allow one to infer, or predict, outcomes which can be observed in employment, and are these the important ones?

"Semi-" or "pre-" vocational courses and assessments are, it follows, those in which validity can be defined partly in this way. It follows that, for example, business studies awards, whether at school, undergraduate or graduate level, ought be validated by reference to more than the course content developed by the relevant teachers. A Masters in Business Administration (MBA) is expected to have some immediate and practical relevance, manifested in the holder's later behaviour in a business environment. Conversely, a law degree which is taken—and designed to be taken—by many students who are not going to become practising lawyers will need to validate its course content partly by reference to legal practice but also partly in other ways.

This defining characteristic of vocational assessment also explains the main concern of those currently occupied with vocational measurement and certification, namely what some measurement experts have dubbed the "criterion problem". The bulk of this chapter will be taken up with this issue and more specifically with the enthusiastically marketed "solution" of competence-based assessment. However, while these validity-related issues dominate discussions in the measurement community, vocational assessment, like educational testing, operates within a wider social and political context, with its own demands and pressures. The final section of the chapter discusses how these pressures affect actual assessment practice, and in particular the effects of social concerns about quality control.

THE CRITERION PROBLEM

We noted above that the distinctive characteristic of vocational education is its reference to (and validation in terms of) outside, occupational activity. None the less, pressures for reliability, costs, time, and the general tendency of large institutions—colleges, vocational examining bodies etc.—to become closed systems all tend to distance vocational assessments and assessors from the practitioner world. Moreover, however highly concerned with validity vocational assessors may be, they are still faced with the problem of finding ways to check it (Wolf and Silver, 1995). In practice, most vocational assessment systems rely on face validity—whether or not something looks right to experts in the field. This, of course, leaves many problems unresolved. Assessments which "look right" may in fact be biased, unreliable, partial in coverage, or weighted in ways which do not correspond to the demands of the occupation concerned.

The problem has increased in direct proportion to the move away from apprenticeship as the dominant mode of vocational training. It is not many decades since apprenticeship was the usual way in which not only carpenters, silversmiths and hairdressers, but also lawyers, architects and accountants learned their trade. The people who were actually going to employ them knew a great deal about their skills and potential long before they qualified; and if standards were not identical from town to town, let alone country to country, this hardly mattered.

With the enormous growth in the scale of formal education and training offered, and in the proportions entering skilled employment, there has been a shift towards provision of vocational education and vocational assessment within specialised educational institutions. This is evident particularly in the case of the professions, where the old forms of articled training have been very largely replaced by university-based training. However, in most countries—Germany, Austria and Switzerland being the main exceptions—similar patterns are evident for craft, manual and office skills. Increasingly, such skills are taught in colleges of further education (UK), community colleges and proprietary vocational schools (USA), vocational and technical high schools (France, Sweden, The Netherlands) or specialised group training associations, supported by consortia of companies (United Kingdom, Italy.) The qualifications issued are designed to have national, and increasingly international, currency. In all developed countries, vocational qualifications are part of a formal system regulated (to varying degrees) by national and/or state law and incorporated into increasingly complex systems of employment and wage legislation and licensing. These changes have, in turn, created new pressures on the assessment procedures used, and generated increasingly formal scrutiny in terms of the concepts of measurement theory.

In this, as in so many other areas of modern assessment research, the experiences of the United States Armed Forces have been seminal: for the quantity and quality of their data, and because of their commitment to formal research and evaluation. It was the American Army's use of mental testing (specifically the Army Alpha test) during the First World War which brought mental testing into the mainstream of American life (Wigdor and Green, 1991); and it was for vocational placement that these tests were devised. In the Second World War, the United States military build-up incorporated from the start formal tests designed to classify and place recruits in training programmes and military specialities suited to their skills and aptitudes.

Particular attention was paid to the selection of aircrew for both the Navy and the Army, with large numbers of psychologists working on the development and refinement of the classification tests. The tests were validated formally, by examining how much more likely high scorers than low scorers were to complete advanced training successfully. On this measure, the tests were extremely successful, with overall completion rates far higher than in the inter-war years, when entry depended on educational qualifications plus a physical alone (Wigdor and Green, 1991; DuBois, 1947). However, the psychologists were also concerned to know whether or not their tests were successful in identifying those who would actually be proficient in combat. Here, the results were far more sobering. The research concluded "with little equivocation that . . . *none of the tests . . . devised to predict success in training, gave evidence of predicting the combat criterion to any marked degree* " (Jenkins, 1950, quoted in Wigdor and Green, 1991, p. 25: my italics)

The experience of the Second World War aircrew selection programme encapsulates the "criterion problem" in vocational assessment. First, as the researchers realised, the tests are only as valuable as they are valid—with the validity defined in terms of later behaviour. Second, measurement of validity is extremely difficult. The Army and Navy researchers could collect data on completion of training quite easily and quite quickly (and very naturally did so.) However, collecting information on combat effectiveness was much harder, much more expensive, and could only be done after a considerable lapse of time. One might add a third point: that when one does manage to collect validation data, they have a habit of showing low correlations with the original assessment (Ghiselli, 1966). While, no doubt, this is in part because of that same lapse of time, it also underlines the difficulty of creating valid vocational assessments—and the likelihood that many of those in use have fairly low validity with respect to the skills and abilities they purport to certify and predict.

In the United States (see Chapter 11) the attention of test developers has been focused overwhelmingly on issues other than validity. Item equivalence, rater error and calibration have all attracted more attention. Wigdor and Green (1991, pp. 28–9 *passim*) argue that

"The full importance of the criterion issue was only slowly recognized by psychologists, and it is safe to say that the problem is largely unappreciated by employers, educators, test takers and others who use or are affected by test scores. In part this is because the statistical assumptions that underlie testing technology are complicated; most people are not conversant with correlational analysis or regression analysis and furthermore do not understand the extent to which the statistical procedures provide the meaning [sic] in this approach to human abilities. In part it is because most people think that test scores have some inherent meaning . . . The failure to fully recognize the seriousness of the criterion problem is also due in part to the sheer difficulty in most settings of coming up with an adequate criterion that is measurable and not prohibitively expensive to develop."

In other words, most of the time, with most vocational and occupational testing, there is little evidence, and fairly little reason, to suppose that scores have any relationship with any relevant "criterion"—that is, with the occupational performance they purport to be about. In the United States, "validity coefficients" may often be cited, but on closer inspection these generally turn out simply to represent correlations with scores on other tests, whose own "validity" is simply taken as given (see, e.g. Chapter 3; Wood, 1991; Blinkhorn and Johnson, 1990). In most other developed countries, validity, as noted above, is simply assumed to derive from the involvement in the testing process of occupational "experts".

In the last few years, however, validity has become an important issue in vocational assessment. There are a number of reasons for this. Some are to do with the shift to certification in formal educational institutions (and the accompanying proliferation of diplomas). Some reflect changes in the demand for labour, and industry and government's increasing concern with developing high levels of skill—and with the high cost of doing so. An American expert on professional evaluation notes this increasing focus on validity in professional-level assessment and argues:

"This is the knotty and persistent criterion problem that for decades has vexed scholars of professional competence evaluation . . . The criterion problem will not go away. It will persist because what professionals think and understand and how they act is far more complicated than what today's assessment technology can probe . . ." (McGaghie, 1993, pp. 239–40).

Another argues that

"The 'problem' resides in the considerable discrepancy that typically exists between our intuitive standards of what criteria of performance should entail and the measures that are currently employed for evaluating such criteria" (Wiggins, 1973, p. 39).

Recommendations for change tend to derive from a position which is both intuitively plausible and supported by a considerable body of evidence (Wolf, 1995); namely that the best predictors of future performance tend to be some sort of "work sample" assessment which accurately mirrors the future behaviour which is of interest. Thus, for example, in large longitudinal studies of British naval officers and senior police officers, their future success could be predicted quite powerfully from training assessments which reflected their future responsibilities (e.g. report writing). Similarly, the record of large companies' assessment centres, which gauge executives' future potential using a range of "work sample" exercises as well as more general ability and aptitude measures is also quite impressive, even allowing for the possibility of self-fulfilling prophecies (i.e. the tendency of companies to promote those whom the centres tip for success) (Wolf and Silver, 1995).

While the idea that "work samples" constitute a valid approach to vocational assessment may seem neither novel nor interesting, their use is, in fact, by no means universal—nor is it unproblematic. In addition to issues of reliability and cost—discussed further in the final section—there are also questions relating to sampling and the extent to which results generalise to other contexts. The single most important development in vocational assessment during the past 20 years has involved the elaboration and development of what is essentially a "work sample" approach, built around the notion of competence. The next section discusses this assessment philosophy and the degree to which it solves the validity or criterion problem.

COMPETENCE-BASED ASSESSMENT

"Competence" as a concept is widely discussed and promoted throughout the developed world (see e.g. CEDEFOP, 1994), but to date it has had the greatest concrete impact in the United Kingdom, where a complete system and "methodology" [sic] of competence-based assessment has been developed and promoted by the government. Comparable systems are also being implemented in Australia and New Zealand, under British influence, and promoted by the United States federal government under the Goals 2000: Educate America Act. Major reforms are also under way in The Netherlands, with the intention of making vocational assessment more competence based; but it is too early for any evaluation data or reports on progress to be available (van den Dool, 1993). The United Kingdom is thus the source of most of the empirical evidence to date on how competence-based assessment operates, and for that reason is discussed in the greatest detail below.

The appeal of the "competence" approach is that it promises a way of developing, guaranteeing and maintaining the validity of vocational assessments—for young people, for firm-based adult training, for professionals and for the low-skilled. Its appeal has been strong in part because so much current vocational assessment seems demonstrably unlikely to be valid, because of its education-based nature, or because it has been tied to the idea of "time-serving" rather than actually learning and displaying skills. There is added appeal in the stated intention of competence-based systems to move beyond the short term. One of the dilemmas for vocational assessment is that it can and must refer to current practice (since that is what is known and is immediately required). Yet all occupations are aware that they face continuing and often massive changes in technology, the work environment, occupational boundaries and the like. Courses, and assessments, based entirely on the immediate, current responsibilities of job-holders are unlikely to predict with any degree of accuracy whether people will be able to cope with the demands of the future.

The objective of competence-based assessment is the assessment of performance or of "practice outcomes" (Curry, Wergin *et al.*, 1993, p. 82). However, the focus is designed to be on the "key purposes of the ... occupation ... and the key functions undertaken" (Fletcher, 1991, p. 167) rather than on detailed analysis of current jobs and routines. The argument is that many assessments have been based on job analysis techniques which focused on too detailed and decontextualised an analysis of what jobs entail, and that this is particularly inappropriate in an age of rapid technological change and rising skill levels. Looking at general "functions" will enable one to see far more clearly what the underlying, general requirements are, on which assessment and certifications should really concentrate. Thus, Fletcher (1991) notes the following may be helpful in distinguishing between these different terms:

> Tasks—activities undertaken at work.
> Functions—the purpose of activities undertaken at work.

OUTCOMES

The most distinctive aspect of competence-based systems is that they start from and remain entirely focused on criteria, or on *outcomes*—the title of the book in which the United Kingdom's chief exponent of the approach set out his philosophy (Jessup, 1991). What does this mean in practice? The first step is an analysis of the outcomes of an occupation or occupational "role". Issues of measurement are not permitted to dominate:—the focus is on purpose and function, not on whether these can be observed, measured or indeed taught. American advocates of this approach to professional recertification argue that "practice outcomes" are the key. This is an answer to the question, "Did the professional achieve the appropriate result?" For physicians, practice outcomes might refer to the end result of medical care: did the patient get better? For teachers it might refer to an end result of the educational process: can the student read? For engineers, it might refer to the integrity of the structure: did the bridge stand? (Norcini and Shea, 1993, p. 82).

In the United Kingdom, this approach has been translated into a complete system of occupational analysis designed to create standards of competence. The first step is a *functional mapping* or *functional analysis* [sic] designed to identify

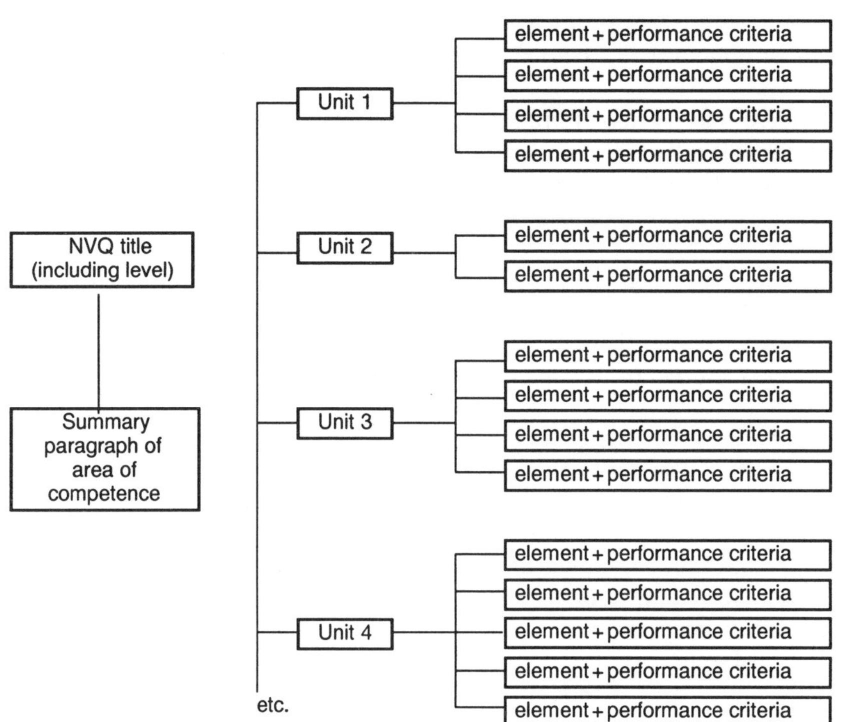

Figure 14.1 The structure of a National Vocational Qualification

"the key purposes of the sectoral occupation and . . . the key functions undertaken" (Fletcher, 1991, pp. 167–8). This is only the first step in actually creating *standards* of competence, however—these being the foundation of a full competence-based qualification under the United Kingdom's system, and of its claims to validity. The next step is to create *units* and *elements* of competence, all expressed in outcome terms, and incorporating parts of the standards; the idea being that each element should correspond to "a discrete function carried out by an individual within an occupational area" (Fletcher, 1991, p. 243). These units and elements are the basis of assessment and accreditation, and reflect the modular structure of competence-based awards. Thus, for example, a National Vocational Qualification or NVQ, as competence-based awards are known, has the structure shown in Figure 14.1.

The heart of a competence-based system, however, is not its modularity, or the idea of "functions", but rather its use of performance criteria. This is where we find the term criterion appearing centre-stage, for these *performance criteria* "reflect the critical aspects of performance—all those qualities which are essential to competent performance" (Fletcher, 1991, p. 169). Examples of performance criteria currently used in United Kingdom qualifications include:

1. Required information is promptly located, obtained and passed to correct person or location (*Business Administration standards relating to filing*).
2. Unexpected events which affect the children are clearly explained and reassurance is given appropriate to their needs (*Child Care and Education standards*).
3. Explained benefits of the product clearly, accurately and concisely (*Retailing standards*).
4. Proposals produced within agreed time-scales and budgets (*Management standards relating to training/human resource development*).

Finally, competence-based advocates often insist that the industry should have *the full responsibility for defining competence*—and, in fact, for recommending how it should be assessed. Nationally established bodies, free and separate from the bodies which actually assess and award qualifications can, it is believed, develop broad national standards which genuinely represent the whole sector. In the United Kingdom, this approach has, up to now, been adopted, so that standards of competence are developed by "lead bodies" representing the occupational sector or industry concerned. The accountants (or rather a group of nominees from large companies and professional associations) develop the accounting standards; the catering industry those for catering; the information technology industry those for information technology, and so on.

Of course, industrial input into vocational qualifications, and the involvement of professionals as assessors is nothing new. However, traditionally—and in most current European systems—industrial representatives advise the education professionals who actually create the award. Competence-based systems give far greater power to industry, on the grounds that this is necessary for the validity of the standards, performance criteria, etc.

Competency-based assessment on a large scale has been implemented to date largely in the United Kingdom, Australia and New Zealand—the major difference being that in the UK the professions have largely resisted adopting it, while in Australia and New Zealand they have generally welcomed it. (This is partly because they have been far less circumscribed in the way they are able to interpret the approach, but also because it has provided an excellent approach to deciding whether to accredit immigrant professionals (Gonczi, 1994)). Its early development, however, was largely at professional level. The basic philosophy of competence-based assessment was first elaborated in the course of American reforms of the 1960s and 1970s, which nowhere involved the craft and manual occupations most associated with the notion of "vocational" assessment.

The impetus for the early United States developments was a growing discontent with the validity of American teacher certification. Influential attacks on the quality of these (Conant, 1963; Koerner, 1963) attracted federal interest, and led to the provision of funds for various reform programmes. Among these were grants to various teacher training institutions to develop model programmes for training and assessing teachers, and it was here that ideas about competence-based assessments were first developed. Similar experimental programmes (on a smaller scale) followed in other areas of higher professional training, most notably the law. The approaches used, and the claims made for their greater validity, were very close to those made in the United Kingdom and Australia during the 1980s and 1990s. However, the American reforms were mostly short-lived and it is via other countries that the approach is now re-entering United States' debate and practice.

THE CLAIM TO VALIDITY

The claim of competence-based approaches to solve the recurrent validity problem in vocational assessment rests on two premises. The first is the process by which the standards are developed, involving direct, and comprehensive, input from employers. An analogy to this approach might appear to be found within the German "Dual System" (of vocational apprenticeship-based training), where national committees of employers, unionists and representatives of the federal government's vocational training agency (the BIBB) thrash out a consensus, industry by industry, which will eventually become national policy. In Germany, however, the emphasis is on the training programme and what this should contain, not on the assessment process. Elsewhere, industry representatives generally advise what are essentially educational committees, rather than *defining* content and standards of achievement. The devolution of the entire "standards development" process to industry is a unique aspect of the competence-based approach.

The second is the nature of the assessment process itself. Gilbert Jessup, for example, the main architect of Britain's NVQs, argues that:

> "Validity . . . implies comparison between the assessments and some external criterion, i.e. that which one is trying to assess. Within the NVQ model of assessment one clearly has an external reference point for assessment—the statement of competence . . . [T]he validity of assessments . . . becomes a matter of comparing the judgements made on the evidence of competence collected, against the elements of competence" (Jessup, 1991, p. 192).

The argument here is that the standards are valid (because of the process by which they were developed) and that they are so clear and transparent that assessment becomes a straightforward matter of comparison:

> "In NVQs the standards of success are already defined and are available to both the assessor and the candidate. Assessment decisions are thus a matter of judgement as to whether the standards have been met" (NCVQ, 1991, p. 21).

The performance criteria in a competence-based system attempt to provide what is, in effect, a definition of "competent" behaviour. If the definition is not clear, then nothing else can be. If the definition is not "right"—if it does not describe clearly the subject-matter or basic "criterion"—then the conclusions people draw about someone's performance will not be accurate or genuinely informative either.

In 1979 a major United States evaluation of competence-based programmes defined competence-based assessment as

> "a form of assessment that is derived from the specification of a set of outcomes; that so clearly states both the outcomes—general and specific—that assessors, students and interested third parties can all make reasonably objective judgements with respect to student achievement or nonachievement of these outcomes, and that certifies student progress on the basis of demonstrated achievement of these outcomes. Assessments are not tied to time served in formal educational settings" (adapted from Grant *et al.*, 1979, p. 5).

As this quotation makes clear, competence-based assessment is essentially a member of the wider family known as "criterion-referenced" assessments. The central importance of the definition process is reflected in all the technical writing on criterion-referencing. The testing literature refers to it as "domain specification" and states the purpose of criterion-referenced assessment as to "ascertain an individual's status with respect to a well-defined behavioral domain" (Berk, 1984). The criterion-referenced literature also advances a position similar to NCVQ's on the issue of assessment itself. "The heart of criterion-referenced measurement is that it provides additional meaning from scores by referencing the test outcome to a clearly specified body of test content, *from which the items are generated* (Millman, 1984: my italics).

These characteristics hold equally for academic criterion-referenced tests, for the early experimental United States programmes and for contemporary United Kingdom competence-based assessments. Indeed, these plus the process of deriving competences on the basis of particular forms of occupational analysis define the approach. They also have an important and demanding implication for the nature of assessment; namely that it be comprehensive, and cover the whole domain, or all the performance criteria. In the vocational or competence-based context, this generally translates into a recommendation that assessment also be continuous. Candidates demonstrate their competence as it is acquired, preferably in a naturalistic setting; and it therefore follows that final examinations are inappropriate. The "portfolio of evidence" (see Chapter 3), showing how much a candidate can already do, and incorporating the assessor's notes, is the hallmark of the approach.

IS VALIDITY ACHIEVED?

The idea of competence-based assessment owes much of its appeal to the pressures of a changing industrial world in which both government and industry feel the need for higher levels of complex skills, and for a workforce which can adapt, "problem solve", and which has "learned to learn" (see, e.g. the preamble to the United States Goals 2000: Educate America Act of 1993). One of the main claims for the approach is that it provides a way of introducing into vocational assessment a broad, forward-looking content, and of escaping from the inherent tendency to be caught in outdated syllabuses, delivered by teachers whose own experience of the occupation is outdated too. This claim rests on the specific notion of "functions" and "functional analysis" and more generally on the allocation to industry of the task of defining exactly what should be assessed for a vocational qualification.

In practice, however, the "standards" which are produced in competency-based systems tend to be voluminous, highly detailed, and to elicit highly atomised assessment approaches, based on tick-lists and checking off boxes. This is just as true for the employer-based lead bodies which are central to the current United Kingdom-developed approach as it was with the earlier efforts of the United States, which were the work of individual reformers based in higher education institutions. (Grant *et al.*, 1979; Wolf, 1995; Callender, 1992)

While lead bodies have certainly made important contributions to defining vocational requirements, their short history also calls into question the concept of "national standards" which can be elicited in a coherent fashion. An unresolved tension was that between the "realistic" need to develop standards which represented common practice and therefore had some hope of being implemented, and the desire to encapsulate best practice, as a way of raising standards and looking to future skills.

With an educational qualification (or one designed to sort and select people) one can, in effect, create a number of different qualifications, by dividing them between different levels. Vocational qualifications, however, tend to be a licence to practise, in which it is important to reach a single set level. This tells the public that one can safely practise, and the employer that one can be employed usefully on the whole range of tasks that may be associated with an occupation—even though the particular job or workplace may use only a subset of these. To this, competence-based systems add the requirement that a candidate must be assessed on, and demonstrably succeed at, every single part of the relevant standards, not simply exceed some pass-mark corresponding to proportional success. Otherwise, one cannot actually know exactly what the "competent" award holder can do; transparency is lost. In this situation, differences between firms in both what they do, and their levels of practice, pose serious problems for the idea of both "national standards" and of workplace assessment.

In both the United Kingdom and Australia it has also proved extremely difficult to get representatives of small or even medium concerns involved, so that standards became largely a product of consultation with large companies (and then often only with a subset of enthusiasts). Finally, as attempts are made to introduce and deliver vocational assessment and awards into the workplace, further tension emerges between a standards development process that aims for inclusiveness and the fact that most companies are highly specialised in their practices and concerns. Many vocational awards simply cannot be assessed in the workplace—the supposed reference point and source of the standards' validity—because companies can only provide a small percentage of the contexts and situations required (McHugh *et al.*, 1993; Crowley-Bainton and Wolf, 1994; Wolf, 1995).

These experiences underline the difficulty of reversing the tendency of vocational education and assessment to become increasingly centred on, and dominated by, educational institutions. The difficulty is greater the more one insists on uniform, national standards. Efforts in France to increase the popularity of apprenticeship have foundered in part on the insistence by the powerful Ministry of Education that apprentices must pass exactly the same examinations as those studying for an award through full-time education. In a country such as Germany, the apprenticeship system has remained strong partly because it is recognised that the "same" training programme and qualification cannot in fact be the same when followed in a small local company or a large, world-class one. However, here too the need to upgrade content and maintain some degree of uniformity is causing strains (Koch and Reuling, 1993).

Actual experience with competence-based assessment also calls into question the second basis of its claim to validity; the nature of the assessment process itself. As we saw above, the argument is that competence-based standards can be so clear that they can be used easily and consistently by large numbers of assessors in different institutions and companies. The clarity of the benchmark means that the same standard can be applied in a wide range of different and authentic contexts.

In practice, however, the more seriously one takes this requirement for clarity, the more tightly the "domains" or "outcomes" or "criteria" tend to be defined, the more narrow and the more numerous they become. The standards developer (or the test developer) becomes caught in an unending spiral of specification: however much detail one provides, however much elaboration one includes, ambiguities still remain.

This process has been very evident in the way competence-based assessment in the United Kingdom has developed. As noted above, what have evolved are standards in which more and more information is included with greater and greater levels of detail, bringing with them atomisation and rigidity. The problem is compounded when, as in most competence-based systems, direct assessment of all outcomes or performance criteria is required. Without such comprehensive assessment, the claim that the assessment or qualification is completely transparent is lost, for one cannot be certain that the candidate has in fact demonstrated all the outcomes and underlying skills. But with it, assessment becomes an overwhelming task, and one on which assessors are almost forced to skimp and cut corners.

The underlying problem is best illustrated with some examples. Figure 14.2 contains a unit from an NVQ at a relatively "low" level, where one might suppose that it is relatively easy to be very specific about what outcomes are expected of a competent individual. It is taken from CaterBase, one of the most thorough and most cited recent attempts to define exactly what is involved in vocational

Cooking food by roasting

ROASTING
Roasting a minimum of two different products, to include: potatoes and a meat or poultry product

PERFORMANCE CRITERIA	NOTES
1 Work area prepared and free from obstruction	
2 Clean equipment of correct type and size assembled and prepared for use	
3 Ingredients of correct quantity, type and quality for amount to be produced are assembled and prepared in required form	
4 Ingredients prepared and assembled in form required for specific dish	
5 Oven and fat preheated and ready for cookery process	• Oven set at high temperature to seal surface initially
6 Appropriate container selected, taking into account quantity of item to be prepared	
7 Seasoning and flavouring added as required	
8 High temperature maintained only for specific products	
9 Even cooking of food without burning, sticking or drying out	• Degree of cooking tested rather than total reliance on time guidelines • Minimum use of fork/skewer to test cooking and to turn
10 Cooked product is of correct appearance	• Food coated with meat juices and roasting fat medium • Skin must not be pierced
11 Product treated appropriately on removal from heat source	• As defined by finishing requirements

HANDLING RESERVATION ENQUIRIES
Handling reservation enquiries, checking for availability and recording and amending reservations

PERFORMANCE CRITERIA	NOTES
1 Response is made promptly	• Must include assessment of response to: • letter enquiry • face to face or telephone enquiry • telex enquiry
2 The correct tariff is quoted including details of service charges, VAT and meals, within individual's authority	• To cover all tariffs relating to the types of room variations, packages offered and pre-negotiated deals
3 Accurate information on room facilities is given on request	• Private bath/shower/toilet/TV/radio/entertainment/tea/coffee-making facilities and room-type variations
4 Accurate information on hotel facilities is given on request	• Food and beverage facilities, including times available • Leisure and sports facilities • Car parking
5 Accurate information on hotel location and local facilities is given on request	• Directions to hotel by car and rail • Local sports and entertainment facilities
6 Enquirer is asked to make a reservation	

Figure 14.2 Sample CaterBase modules

B113 Agree a strategic plan which meets organisational requirements

(a) Proposals are presented which match the implementation of agreed strategy to organisational requirements and priorities
(b) Proposals identify accurately when outcomes critical to organisational requirements are to be achieved
(c) Responsibilities for achievement of and support for the plan are clearly defined
(d) Proposals are based on agreed organisational development and resourcing strategies and plans
(e) Proposals are produced within agreed time-scales and budgets
(f) Decision makers are given adequate opportunities to ask questions and seek clarification
(g) Negotiations and agreements are conducted and concluded in a manner which preserves goodwill and trust
(h) Agreed plans are summarised accurately and made available to those who require this information.

Range Indicators

Types of plan:	written/oral; free standing/part of a corporate planning system
Relationship with decision makers:	peers/subordinates/superiors: employers/employees/clients
Presentation methods:	oral, written, audio-visual, computer-based
Presentation/negotiation processes:	direct/face to face: remote/via technology based systems

Figure 14.3 Element for a higher level NVQ (from Training & Development Lead Body Standards)

"competence". The purpose here was to lay out performance criteria which reflected precisely industry requirements and consensus, and to do so in such an unambiguous way that CaterBase modules could be assessed and accredited by workplace supervisors throughout the huge catering industry. And indeed this is happening, with large parts of the assessment in current catering NVQs the responsibility of workplace assessors who are expected to use CaterBase requirements as a definitive guide to the standards of performance expected.

Yet even here it is only too easy to imagine confusion over what a competent person has actually done or can do. Suppose we take a single performance criterion from Figure 14.2: "Ingredients of correct quantity, type and quality for amount to be produced are assembled and prepared in required form" (no. 3). The nature and difficulty of this task is going to be very different if the context is a small kitchen in a country inn, or a huge city hotel preparing enormous joints for a buffet. Preparing the ingredients for small stuffed partridge or guinea fowl will be quite different from what is required for a single unstuffed chicken or small leg of lamb. How does one judge if the candidate has produced ingredients of "correct quality"? Can they be poor quality if the candidate is working in a rather unsatisfactory restaurant, and if so, how is a future employer to interpret the statement that the candidate met this criterion?

One can repeat this process of endless interpretation and call for more clarification and detail with every one of the performance criteria in Figure 14.2. Nor is this a problem specific to a certain vocational area, or level of award. Figure 14.3, which contains an element from a much higher level NVQ, intended for a training manager in a company setting, underlines the problem. How do we compare the individual who produces proposals "within agreed time-scales and budgets", and who is working in an environment with adequate funding, good support staff, and realistic managers, with someone who enjoys none of these? How many decision-makers are meant to be "given adequate opportunities to ask questions"? How complex, or simple, is the plan in question meant to be?

There are some general lessons here, relating to criterion-referenced measurement of all sorts, whether vocational or academic in its terms of reference (Wolf, 1993). However, there are also some more specific issues which relate directly to vocational education, and to whether competence-based assessment can in fact satisfy the demands of students, employers, and the wider society.

The first is that one cannot, in fact, ignore measurement issues. The fact that assessments which derive essentially from classical measurement theory are so inadequate in a vocational context does not mean that simply starting from the "valid"—the analysis of occupational functions—is a satisfactory substitute. Can one really, following the advice of competence enthusiasts such as those quoted above, accredit doctors on the basis of whether a patient gets well? or an engineer on whether a bridge stands?

Even when developed fully, by occupational experts, analysis of what "competence" involves tends to leave one well short of what is feasible, let alone practicable, in assessment terms. As demonstrated above (Figures 14.2 and 14.3), this approach leads to many outcomes which *simply cannot be measured*. It is, in practice, impossible to tell whether "proposals identify accurately when outcomes critical to organisational requirements are to be achieved" (Figure 14.3); or whether "indicators of contingencies/problems" have all been "referred to an appropriate authority".

The result, in practice, is that they are not measured at all. The risk attendant upon a competence-based approach is thus that one will obtain, on the one hand, a system overloaded with atomised assessment criteria, dealing with the short term and immediate, and on the other, an array of outcomes which are broader, more concerned with general functions and future developments, but not actually part of the assessment process. Describing comparable experiences in nursing, where practical, work-based assessment has been increasingly emphasised because of its supposed greater validity (albeit without a full competence-based system), an experienced nurse-educator observes that

> "with . . . increasing (work) pressures . . . it could be argued that continuous practical assessment is in great danger of becoming no assessment at all. The extent to which experienced nurses are able to supervise and give the continued feedback on learners' progress that was considered such a significant improvement on the previous system of assessments is questioned . . ." (Girot, 1993, p. 89).

Problems such as these have led to criticisms from observers in a number of different countries. A recent United States review of existing occupational (or vocational) "performance assessment" systems concludes that the competency-based systems developed to date have, as actually implemented, produced assessments of a very traditional sort, in spite of the claims of their advocates that they represented a breakthrough in vocational assessment. Wirt (1995) sees the current United Kingdom/Australasian model as no different in essence from what he calls the "job competency model" underlying most job analysis in industry, and short-term occupational training in industry and military contexts. Under this approach,

> "Skills are defined using a job task inventory to identify the specific tasks that people perform in their jobs. Assessments are conducted by having supervisors or trained personnel observe job performance and "check off" task-by-task which tasks a person can perform at what general level and which the person cannot perform. This method evolved from the techniques of organizing production introduced around the turn of the century by Frederick Taylor" (Wirt, 1995, p. IV-1).

As noted above, the United States has committed itself to developing new "industry skill standards", designed to increase skill levels for a world which its government perceives as increasingly competitive and for which it judges existing education and training to be inadequate. Skill standards and competence-based assessment were certainly not intended to perpetuate a Taylorian approach. On the contrary, it was quite explicitly intended not to tie vocational qualifications to the specific requirements of existing jobs, but to go beyond detailed task and job analysis to "functions" and occupational competence broadly defined. Yet it is Wirt's judgement that these standards are, in fact, likely to reflect nothing but the "job competency model" writ large, and much United Kingdom and United States experience is consistent with this.

It follows that the "validity issue" is not so easily solved as the competence advocates believe. Analysing competences is not enough if one cannot actually assess them directly, either because the outcome statements are essentially non-assessable or because of the simple volume of discrete tasks generated. (This latter problem effectively killed off all the earlier United States attempts.) In that case one is faced, again, with the question of whether "proxy" measures, or samples of behaviour, are acceptable, and how one is to assess their validity. It also becomes obvious why, in practice, so many vocational assessment systems have substituted periods of experience for direct measurement of any sort at all. If—as is often the case—there simply is no classroom-based substitute for the lessons of the workplace, and if, in practice, any formal and objective assessment during work is impractical, substituting the requirement for a "period of experience" may actually be the rational strategy.

Critics of recent "competence-based" reforms do not, on the whole, advocate a more educational approach, or a rejection of performance assessment and "job samples" in favour of psychological tests of underlying abilities. The latter remain discredited among many policy-makers and measurement experts, and instead what is called for is better, more sophisticated performance measures and more "lifelike evaluation approaches" (McGaghie, 1993). But we have very little idea of exactly what it is that we are trying to achieve by way of such measures, or where to find exemplars. In the meantime, as the next section discusses, there remain important pressures on vocational assessment systems over and above the resolution of the "criterion problem".

QUALITY CONTROL

The typical approach to vocational assessment—whether we are discussing plasterers, electricians, lawyers or surgeons—has been to combine "continuous" assessment with some sort of "end test". This was the approach of the medieval guilds, where apprentices served time under a master but also had to present a final "masterpiece" to qualify, and it finds direct parallels in the training of solicitors serving articles, or chartered engineers or doctors who must complete so many years of practice before obtaining their final qualification.

Apprenticeships generally retain this pattern. Although in most of the developed world only a very small proportion of young people enter formal apprenticeship, it remains a major part of post-compulsory education in Germany, Austria and Switzerland. The major guarantor of a "graduating" apprentice's skills is their successful completion of the process of training, with its highly specified content. Nonetheless, it is also necessary for all apprentices successfully to complete final tests. Under the German system these combine long practicals taken under examination conditions with oral questions from the examining committee (employers and skilled workers appointed by the local Chamber of Commerce) and written knowledge tests (predominantly purchased from one of the agencies which specialises in producing these at a national level).

Vocational qualifications delivered by educational establishments follow a not dissimilar pattern in other parts of Europe, although the importance of the written element is generally greater, and outside the United Kingdom there is usually a heavier weighting, within the course, towards "general" education at the expense of specific occupational skills. Nonetheless, the pattern is of "end tests" which combine written and practical elements, and some continuous assessment and/or a requirement that certain courses be followed before the qualification can be obtained. This commitment to "time serving" is one of the aspects of conventional vocational assessment systems to which "competence" advocates object, seeing it as an artificial barrier to qualification.

The major exception to this pattern is the United States, where the licence to practice in many occupations and professions is in the gift of state governments and tied to licensing examinations or tests which are separate from training course assessments. While some (e.g. those for lawyers wishing to practise in a state) are developed and operated separately state by state, many others are developed and offered nationally by either one of the big testing companies such as Educational Testing Service (ETS) or by specific organisations or boards. The most distinctive aspect of these examinations is the number which are multiple-choice tests developed by measurement experts within the psychometric/classical test theory tradition (see Chapter 11). Enormous care is devoted to item analysis, to studies of score distribution and methods of standardisation to ensure equivalence year by year in the difficulty of the tests and to statistical modelling (see Chapter 4). Another distinctive feature compared to common practice elsewhere in the world is that candidates are generally free to take the tests repeatedly.

An overview of current practice also makes it clear that there has been rather little change in recent years outside the few countries which have experimented with competence-based assessment systems. The tying of licensure to multiple-choice tests was the issue which engaged (and enraged) the United States reformers of the 1960s, and led to the first competence-based experiments. But while these originally appeared likely to change the whole nature of teacher education, qualifying as a teacher in the United States today is again largely a question of fulfilling requirements to take courses and passing a standardised test. The other experimental programmes in the United States have vanished entirely. The current widespread calls for reform, and for assessment which is more valid and related to actual practice, are similarly directed towards the dominant model of licensing on the basis of multiple-choice psychometric tests. While many of those involved in the development of such tests, however, admit to the force of the arguments, there is very little sign of change in practice (McGaghie, 1993).

In European countries other than the United Kingdom, interest in the idea of "competence" and in the implications for vocational education of a changing world economy are seen as having as much or more to do with the curriculum as with the mode of assessment. There is considerable teacher resistance to any increase in continuous assessment and no apparent move to jettison final examinations in line with the United Kingdom's experiment.

The apparent discontinuity between the experts' recommendations and actual practice finds a simple explanation in another of the organising concepts of assessment: *reliability*. Even with the use of practical end-tests, carried out under examination conditions, the European model (shared by the United Kingdom and Australia until the recent reforms) makes it difficult to ensure reliability of marking and cross-site comparability of standards. Vocational assessments are relatively less likely to use paper and pencil tests, of a type which can be set and marked centrally, and more likely to use practical tests which cannot, by their nature, be collected together, or marked centrally, or kept for long periods of time. (This is not to say that the vocational and the practical are synonymous. Physics assessments are frequently practical: actuaries' examinations are paper-based.) It is therefore difficult and very expensive to monitor standards—or train and check the consistency of assessors.

Conventional academic examinations of the European type, and the type of paper-based "performance assessments" advocated by many assessment experts, all allow the possibility of conventional quality control measures such as remarking, examiners' meetings, and the like (see Chapters 5 and 7). The products of much vocational assessment are intrinsically ephemeral (e.g. a doctor's examination and verbal diagnosis, a selection of cakes) or impossible to transport in bulk (e.g. a carpentry or plasterwork project). Moreover, they tend to have relatively large numbers of assessors—and far more who are not "professional" assessors. In order to promote both relevant, workplace experience, and ensure that up-to-date standards and judgements are incorporated into the assessment process, even college-based courses often include compulsory workplace components, and it is very common to use practitioners as sole or joint examiners. Thus, in the United Kingdom, the Inns of Court use practising lawyers to assess the compulsory skills course for aspiring barristers, while in Germany the Chambers of Commerce which oversee the final assessment of apprentices depend on the (unpaid) services of committees of qualified workers and employers.

While such practices are not only helpful but necessary if vocational assessments are to remain closely tied to occupational practice, they inevitably create worries about how far assessors are using common standards, and following good assessment practice. Research on performance assessment in academic fields underlines the huge amount of variability in scoring which results from differences among tasks. As much as half the variation in students' scores is ascribed, in recent studies, to differences among the supposedly equivalent tasks performed within the same assessment domain (Shavelson *et al.*, 1993; Dunbar *et al.*, 1991). Compared to this, the effect of variations in the way assessors scored and rated was very small. Yet huge variability in the nature and context of tasks is even more apparent in the different realistic setting of "performance assessment" in vocational settings than in the context of an academic programme. This is not a situation likely to reassure the trustees of a licensing board.

Concerns over assessor reliability and comparability of standards become increasingly important the more qualifications become a national and indeed international currency: exactly the situation at present. While assessment experts are most occupied with issues relating to "the criterion problem", politicians, industry and the general public tend to be far more concerned with reliability. An ambitious attempt by the European Union to establish comparability between qualifications, in order to secure mutual recognition, was carried out on the basis of course content and assessment objectives. It foundered completely on being presented to national bodies (notably the German employer and labour organisations), who were not prepared to accept this paper-based analysis as evidence that German standards actually were being achieved in other countries.

As discussed at some length in the previous section, the country which has, in recent years, promoted validity at the expense of reliability is the United Kingdom. (The trade-off is expressed in just such terms by the main proponents of a competence-based approach, who argue that "reliability is not an issue" in the context of performance assessment against standards (Burke and Jessup, 1990, p. 195)). However, there is increasing anxiety and public criticism of the NVQ system on reliability grounds (see especially Prais, 1991).

While there have not as yet been any moves to introduce end-tests or other centralised assessments into NVQs, the recent introduction into England of semi-vocational, education-based General National Vocational Qualifications (GNVQs) indicates the importance of reliability concerns at political level. Although the new GNVQs are supposedly "competence-based", the Prime Minister made a personal decision, against the advice and arguments of the National Council developing the awards, that all GNVQs must have a major element of centrally set, centrally marked external assessment.

For American vocational assessors, reliability is even more of an issue. Vocational and professional tests are "high-stakes" affairs—the more so the more they are tied to licensing. They also operate in an environment where legal challenges are a genuine possibility, and a daily concern to test developers. A personal anecdote will illustrate this point. At the very beginning of the competency-based movement in the United Kingdom, I gave a seminar on the subject to a group of Washington-based experts, explaining how competency-based assessment, using workplace assessors, was to be the basis of nationally delivered, nationally recognised qualifications, not simply an approach taken up—like the early United States experiments—by individual universities for their own degrees. Their immediate response was incredulity. How, they asked, could the British possibly combine such a decentralised, non-standardised system with licensure requirements, and maintain this in the face of court challenge?

The more important and the more legally based the qualification, the more reliability and "fairness" or equity outweigh considerations of content and relevance. And the more important equity becomes, the more assessors are thrown back onto paper-based, and "standardised", techniques. Such approaches can demonstrate their reliability; and they have the additional advantage of relative cheapness. Moreover, a growing body of United States judgements and opinions confirms that the important issues in deciding whether a test is legitimate as a basis for licensing or qualification are (a) whether or not it can be shown to be related to the curriculum in content terms, and (b) whether it follows the rules for test development laid down by the (psychometrically inspired) Standards. Multiple-choice standardised tests regularly meet these courtroom criteria.

The United States encapsulates in more extreme form a recurring tension in vocational assessment, for which quality control poses problems quite different in scale from those faced by academic assessors. The reason that competence-based approaches have attracted so much interest in recent years is that they promised to reconcile the demands of validity with those of quality control—and to solve the cost issue by pushing large amounts of assessment into the workplace. The experiences of those countries which have experimented widely with competence-based systems suggest that the optimism was premature. The problems and the tensions remain.

CHAPTER 15

Assessment in the Workplace

Roderic Vincent

Human Qualities, London, England

INTRODUCTION

Organisations increasingly see people as the main determining factor in their success or failure. In an age of flatter organisational structures, with fewer people doing more important tasks, and with greater financial accountability for managers, and emphasis on flexible teamworking, customer service and quality at all staff levels, organisations are eager to know whether they are selecting the right people and how well their existing people are performing. As a result, staff are assessed frequently and in many ways. Some typical forms of assessment include selection for jobs, promotion decisions, performance appraisal and the identification of training needs. In addition to this, assessment is carried out daily as an informal process by managers and supervisors. The rigour of workplace assessments varies from the completely informal to sophisticated systems set up at great cost by large corporations. At the more systematic end of occupational assessment we tend to find selection and assessment systems which include formal techniques such as psychometric tests and questionnaires, and assessment and development centres. While recognising the importance of other more informal types of assessment, this chapter will concentrate mainly on the formal methods, and the issues and problems that face the occupational psychologists and human resources professionals who apply these methods in the workplace.

This chapter reviews the current issues surrounding the major assessment methods in use in organisations, including ability tests, personality measurement, biodata, assessment centres, interviewing and work sampling. These methods are fairly widely used in many countries, but differences do exist between nations. For example, general intelligence tests are more common in North America than in the UK, where specific ability tests have tended to be popular (Muchinsky, 1994). Integrity or honesty testing is also less common in the UK. Even within the European Union there are distinct national differences in the assessment methods used by organisations. In France, for example, graphology is more popular, and assessment centres are more prevalent in Germany and the UK

Assessment: Problems, Developments and Statistical Issues.
Edited by H. Goldstein and T. Lewis.

than in Italy, where the interview tends to predominate. These differences across nations may result in some cases from different legislation, such as the stricter legal constraints under which US corporations operate. Shackleton and Newell (1994) also suggest that different research traditions and cultural differences may have contributed to these variations in practice.

Without doubt there has been an explosion in the use of assessment techniques in the workplace in recent times. Testing and assessment are big business with an enormous growth in the products available. The fourth edition of *Tests in Print* (Murphy *et al.*, 1994) which provides a comprehensive list of commercially available tests includes 3009 entries. This is 22% more than the previous 1983 edition and 43% more than the first edition (1961). In the UK, a survey of test use for management selection in the top 1000 companies (Shackleton and Newell, 1991), found that as many as 41% now use cognitive tests. However, a survey conducted at the same time that included smaller companies (Smith and Abrahamsen, 1994), found only 5.5% were using tests. The subject-matter of this chapter, therefore, concerns assessment methods that are used almost exclusively by the larger companies, those which can afford the resources needed to research, set up and run elaborate assessment procedures. If, as predicted by forecasters such as Naisbitt (1994), it is the small- and medium-sized companies that will be most important to the global economy in the future, then it will be essential that they have access to the most effective forms of assessment in the workplace. Ways in which this might be made possible are discussed in the final part of the chapter which looks at future developments in occupational assessment.

WHAT MAKES OCCUPATIONAL ASSESSMENT DIFFERENT?

Assessment in the workplace presents many of the same problems as educational assessment, discussed in other chapters of this book, with particular parallels to vocational assessment discussed in Chapter 14. Occupational assessment also has its own distinct problems. Some of these stem from the diverse nature of organisations and jobs, and others from the commercial constraints of the world of employment. Every organisation is unique, and some are fiercely proud of their individual identities. Workplace assessment needs to fit in with the culture of the organisation or it will be rejected by employees and management (Handy, 1986).

Some organisations have a scientific and rationalist approach to management and are likely to respond positively to the idea of conducting job analysis or validity studies. Others operate at a more intuitive level, or would see any systematic approach to assessment as too bureaucratic for their style. Assessment also needs to be acceptable to the assessees in organisations, in a way that is often less important in education. For example, an occasional response from managers asked to undergo assessment is "Why do you need to use these tests? Surely the organisation knows what it thinks of me by now." This type of reaction would be less likely and probably easier to respond to from a student asked to sit an examination. In organisations, it is often deemed inappropriate to assess the most senior people, as this would be seen as demeaning to their status.

Once assessment has been integrated into an organisation it can become part of the rituals and language which make up the culture (Mitroff and Kilman, 1976; Deal and Kennedy, 1982; Trice and Beyer, 1984). This can be seen in large organisations committed to assessment such as British-American Tobacco where the competencies used for assessment are well known throughout operating companies across the world, and where managers can be heard using abbreviated forms of the names such as "RDM" for "Rational Decision Making". Once the initial barriers and fear of the unknown have been overcome, assessment can take a strong root in many organisations.

The diverse nature of organisations and jobs also leads to practical problems. In education there are frequently thousands of students undergoing essentially the same educational experience. This allows researchers access to large sample sizes, for example, when calibrating tests against each other, evaluating the effectiveness of assessments, or trying to establish the fairness of tests across different groups. In the workplace, however, many jobs have only a few incumbents. The most important positions, from the point of view of selection decisions, are held by only one person. This makes job analysis, to determine the criteria for success, difficult, and makes large-scale criterion-related validity studies impossible, although to some extent this problem has been overcome by the use of meta-analytical techniques to combine the results from many small-scale studies. The results of these meta-analyses are described later in the discussions of each of the primary selection methods in use. Some assessment methods such as biodata, discussed later in the chapter, require large numbers of people doing the same jobs for their development. This has limited their spread in occupational use, and made them more acceptable to certain types of organisation such as the armed forces.

Occupational testing also has some advantages over educational testing. In a recent comparison of the two fields, Camara and Brown (1995) point out that performance criteria may be easier to define in industry where there are specific organisational goals such as profit and productivity, compared with the broader and possibly more vague goals of education. They list several examples of harmful and unintended consequences of educational assessment. Some of these, such as overemphasis on rote memory and exam content, have not troubled those concerned with employment testing. In other examples, such as allowing test results to serve as the sole basis of decision making, employers have tended to use a wider variety of assessments, including multiple assessment procedures for selection, and continuing performance management for existing employees.

One final example from Camara and Brown (1995) that is worth noting is the phenomenon of teaching to the test. The problem of teachers devoting time to test content rather than to relevant curriculum beyond that sampled in the test has not traditionally been mirrored in occupational assessment. However, the recent fixation by employers on competencies (see Chapter 14), could result in staff expending more effort on demonstrating that they have reached the required competency levels, rather than ensuring that they contribute to the organisational goals, especially if competency frameworks are not closely related to organisational goals, and frequently adapted as the business needs change. Teaching to the test could also become more problematic in industrial settings following recent moves towards the accreditation of training by government or educational establishments. This is perhaps where assessment in the workplace meets educational and vocational assessment.

THE MAJOR ASSESSMENT METHODS

We now look at the main types of occupational assessment methods.

Ability tests: prediction and social costs

An enormous range of tests of aptitude and ability exist, covering everything from visual acuity and manual dexterity to cognitive tests of verbal, numerical, and spatial ability. There are also many tests of job specific skills such as mechanical reasoning or computer aptitude. Occupational test developers have tended to produce ability-specific or even job-specific tests rather than using general measures such as *g*, Spearman's concept of general ability or intelligence. They have argued that validity is increased when the content of the test closely matches the content of the job. In other words content validity has been considered important for occupational test choice.

However, recent research into validity generalisation has questioned this view, and raised a number of other issues that are highly topical in the field of occupational testing. The research using meta-analytical techniques to combine the results of many studies has indicated that general ability tests show strong predictive powers across all job types (Schmidt and Hunter, 1977; Hunter and Hunter, 1984). These methods and their results can be seen as an answer to the problem of conducting validity studies, which would normally require large samples of employees, in settings where there are only a few job holders. They could also be interpreted by the time-pressured personnel manager as an argument for using any well developed test, obviating the necessity for job analysis to determine the exact skills required for success in a particular position. However, interpreting the implications of the results of validity generalisation is controversial. As Cronbach (1990) puts it:

"The Schmidt-Hunter position runs counter to the long-standing American interest in the diversity of aptitudes and to the undeniable fact that jobs within a category—even jobs with the same name—differ in the talents they demand."

The case for validity generalisation also raises questions with regard to the fairness of ability tests (Schmitt and Noe, 1986; Schmidt *et al.*, 1992). Although tests of general ability have been shown to be valid predictors of performance across a wide range of jobs, and to predict equally well for various ethnic groups (Hunter *et al.*, 1979), the tests have also been shown to differentiate between ethnic groups. This could lead to selection decisions being made in favour of white candidates at the expense of others. According to Schmidt *et al.* (1992):

> "Failure to use such tests would cause great productivity losses to employers and the economy, losses we can ill afford in today's competitive international economy. But increasing use of such tests could mean fewer minorities hired in the more desirable jobs, or at least could make it difficult to increase current minority representation."

There is therefore a social dilemma associated with the use of tests, even when their criterion-related validity has been established (see Chapter 6). A number of approaches have been tried to reduce the problems of using ability tests with diverse groups. These have included providing practice materials, using practice tests, and setting low cut-off scores. Some success has been reported with these approaches. British Rail, for example, reviewed their testing procedures including looking at access training and practice tests following a court case for alleged racial discrimination against train guards (Callen and Geary, 1994; Kellett *et al.*, 1994). More needs to be known about the effects on validity and utility of these solutions to the problems of adverse impact. Hunter and Hunter (1984) suggest that using low cut-off scores, for example, can have dramatic effects in reducing the validity of tests. An optimistic view is that we can adapt testing procedures to ensure that adverse impact on different ethnic groups is minimised, without serious loss of validity or utility. A pessimistic view would be that employers and society in general will have to choose between the competitive advantage afforded by ability tests, and the need to ensure that diversity is managed and selection systems are fair.

In order to demonstrate fairness in the tests they use, test publishers and employers need to provide evidence of content and construct validity, rather than relying solely on the results of meta-analysis studies demonstrating that the tests are generally valid in a variety of settings. This means, for example, that thorough job analyses should be conducted to identify that the constructs measured by the test are reflected in the abilities needed for success in the job. It is important to demonstrate not only that the test is sound, but that it is the right test, and being used correctly.

Personality measures: are they valid?

Personality questionnaires have a long history of use as an employment assessment method. They have also long been questioned as valid predictors of job performance. Thirty years ago, Guion and Gottier (1965) concluded a major review of personality measures in personnel selection with the statement:

"it is difficult in the face of this summary to advocate, with a clear conscience, the use of personality measures in most situations as a basis for making employment decisions."

Despite repeated scepticism by researchers, personality questionnaires have been popular with practitioners. Their appeal may lie to some extent in the belief that they provide insights into characteristics that are crucial for success at work, but which are difficult to gauge at interview. Those responsible for selection in organisations may believe that a candidate's mental abilities can to some extent be judged from an educational track record, so ability tests may not appear to reveal too much. Attributes such as creativity and resilience under pressure which personality tests purport to disclose are greatly appealing to the manager trying to make important selection decisions.

Recently, there has been a resurgence of debate about the validity of personality measures, following meta-analytical reviews of the available validation studies (Barrick and Mount, 1991; Tett *et al.*, 1991; Robertson and Kinder, 1993). Meta-analyses have indicated that some of the disappointingly low correlations found between personality scales and job performance measures with small sample studies are in fact more impressive when the data are combined to give larger sample sizes. Although some of these studies, particularly Tett *et al.* (1991), have been criticised on methodological grounds (Johnson and Blinkhorn, 1994; Ones *et al.*, 1994), it seems we are moving slowly towards a position where the majority of organisational psychologists believe there is enough evidence to justify the continued use of personality questionnaires in personnel decision-making.

Given that position, the emphasis should be on making appropriate use of appropriate questionnaires. This means collecting thorough information on the job demands to establish exactly what personality characteristics are required for success, selecting psychometrically sound tests, conducting local validity studies with *a priori* hypotheses about which dimensions will be related to the characteristics, and using the questionnaires in conjunction with other valid assessment methods rather than in isolation. It also means ensuring that the tests are administered by appropriately trained people, and that those who interpret and give feedback on the results are experienced with the instruments, familiar with their limitations and with current research findings.

Having decided that personality questionnaires can be valid in some circumstances, one of the decisions that faces the practitioner is whether to use a questionnaire that measures a wide range of traits, or sticks to the "Big Five" scales that have repeatedly arisen from factor analyses of personality dimensions (Digman, 1990). Where researchers have used factor analysis to summarise the data from personality dimensions, more or less the same five factor solution seems to pop up. The factors listed in Table 15.1 are given different names by different researchers, but seem to provide the most reliable structure of personality from an assessment point of view.

The Big Five have generated a large volume of research in personnel psychology (e.g. Schmit and Ryan, 1993; Mount *et al.*, 1994), and have provided a

Table 15.1 The Big Five personality dimensions

Factor	Other names	Description
Extroversion	Surgency.	The tendency to be sociable, assertive and talkative. This has sometimes been seen to include ambition.
Emotional stability	Neuroticism, Stability, Emotionality.	The extent of anxiety, insecurity, worrying and the tendency to embarrassment.
Agreeableness	Likeability, Friendliness, Social Conformity, Compliance, Love.	The tendency to be tolerant of others, good-natured, co-operative forgiving and soft-hearted.
Conscientiousness	Dependability, Conscience, Will, Conformity.	The tendency to plan and organise, be thorough, responsible and careful. This has also sometimes been associated with hard-working, the need to achieve, and perseverance.
Openness to experience	Intellect, Culture.	The tendency to be broad-minded, intellectual, curious, imaginative and artistically sensitive.

focus for looking at the validity of personality questionnaires via meta-analysis. Barrick and Mount (1991), for example, found that the "conscientiousness" factor was consistently related to job performance across a wide range of jobs from the police to salespeople and managers.

For the manager who wants to include personality measurement as part of an assessment procedure, the choice of whether to use a Big Five questionnaire or one that measures 10, 20 or more traits, is likely to be based on practical needs. The Big Five may be the most elegant statistical solution, but will not always give enough detailed information of the type that is needed in assessment. The dimensions are very broad; for example, looking at the conscientiousness scale we might want a senior manager to be hard-working and achievement oriented, but not too detailed in work style. Equally, from the openness to experience factor, broadmindedness and intellect might be important, but not artistic sensitivity. On the other hand, questionnaires that contain a wider range of scales are likely to overlap in the information produced, with some scales providing very similar information to others. Once again, it is a matter of the sensible application of the instruments. Following systematic consideration of the job requirements in the context of organisational goals, it should be possible to decide between the summary that can be provided by a Big Five approach or the greater detail, and greater overlap, from questionnaires with more scales.

Biodata: a neglected area

Biodata are a more systematic method of collecting the sort of information that appears on application forms. Biodata cover scorable information from people's experience, and have previously included demographic, experiential and attitudinal items which are demonstrated to be relevant to job performance. Originally, application forms were converted into weighted application blanks which are invisible to candidates in the sense that they appear to be normal application forms, but contain questions that have undergone validation and can be scored. More recent biodata forms have tended to use multiple choice formats which facilitate more accurate and rapid scoring, and allow for optical mark reading machines to score large volumes of applications very quickly.

Mael (1991) provides a useful taxonomy of biodata items, highlighting the type of items that have been shown to work as predictors of job performance. Biodata should be historical rather than hypothetical. Historical items which are, at least in principle, verifiable encourage honest reporting by candidates, and are seen by Mael (1991) as defining biodata (e.g. "How old were you when you got your first paid job?"). Hypothetical items such as "Where do you want to be in five years time?" which do not cover factual data, produce less reliable information, and so should be excluded from biodata questionnaires. Other examples of the characteristics of successful biodata items are that they should be external rather than internal, that is dealing with events rather than attitudes and opinions as in personality questionnaires. They should be objective (e.g. "How many times have you been late for work in the last year?") rather than subjective ("Would you describe yourself as punctual?"). Importantly, for fairness, biodata items should be controllable by the candidate (e.g. "How many attempts did it take for you to pass your accountancy exam?") rather than non-controllable (e.g. "How many brothers and sisters do you have?"). To ensure fairness, they should also be evaluated to ensure that all groups have equal access to the types of activities described.

Once the characteristics above, together with others listed by Mael (1991), have been met, biodata can be powerful predictors of job performance. Gunter *et al.* (1993) in their book on biodata, recount the case of the single item "Did you ever build a model airplane that flew?", which apparently was almost as good a predictor of success in flight training during the Second World War as the entire US Air Force Test Battery. In meta-analytical reviews of validity, biodata have compared very favourably with other assessment methods. Hunter and Hunter (1984) found that for entry-level jobs biodata had an average validity coefficient of 0.37 compared with 0.53 for ability tests and 0.14 for interviews when the results are correlated with job performance measures. This is impressive given the simplicity of collecting biodata information compared with the other assessment methods. Biodata forms can be sent out as application forms rather than having to be administered under strictly standardised conditions as for ability tests, and do not require the expense of interviewer time for their interpretation.

Given their validity, it is surprising that the use of biodata is not more widespread. This probably stems from the practical problems associated with the development of biodata. Large sample sizes are required for item trialling and validation. Cross-validation is important, where the results from one group are tested on another to eliminate the chance effects to which biodata are prone. For example, in the recent development of biodata for graduate pre-screening at British Steel, over 200 recent graduate entrants took part in the initial validation, and a further 200 were used for the cross-validation. Many organisations would not have access to such large numbers doing the same or similar jobs for the development of biodata questionnaires.

Rothstein *et al.* (1990) provide further evidence for the validity of biodata, and some hope for its extended use. They have found, contrary to the prevailing view that biodata items are highly situation specific, that there is some evidence for the generalisation of biodata validity across organisations. They conclude by suggesting that this may provide a way forward for biodata through consortium-based, multiple-organisation biodata research. Organisations with smaller numbers of employees doing similar jobs could pool the costs and the results from the development of biodata questionnaires to avoid the need for a single large sample.

Assessment and development centres

Assessment centres combine data from a number of methods such as psychometric tests, job simulation exercises and interviews, allowing assessors to observe candidates in various situations, such as group exercises, role-plays, presentations, and so on. They use exercises that are closely based on the tasks performed in the job, bring together the best aspects of several assessment methods, and provide alternative viewpoints on candidates, usually from several assessors, all of which should help to reduce the likelihood of error. They have consequently been shown to have strong validity in occupational assessment (Hunter *et al.*, 1982, Gaugler *et al.*, 1987). The use of assessment centres has grown rapidly in recent years, and is now widespread. A survey by Boyle *et al.* (1993) of usage in UK organisations employing more than 1000 people found that over 45% make some use of assessment centres. As Adler (1987) aptly points out,

> "In contrast with many other developments in human resources management over the years, this widespread adoption of assessment center technology was initially stimulated not by uncritical enthusiasm for the latest personnel fad, but by the impressive results of sound empirical research."

In their early days assessment centres were used mainly for selection, primarily management selection. This quickly spread to use with other jobs and extended to many types of application including promotion decisions, the identification of long-term potential, diagnosis of training needs, evaluating the effectiveness of training, and so on.

The moves towards the use of assessment centres in connection with training and development have led to the centres frequently being termed development centres, and this overlap between assessment and development in occupational assessment is reminiscent of the discussion of grading versus diagnosis in Chapter 1. In some cases the developmental benefits of the centres have been assumed simply from the fact that participants were given feedback following the assessment, with no changes made to the content or process of assessment of the exercises. The assumption is that the experience of attending a centre and receiving feedback is developmental because of the increased self-awareness of strengths and weaknesses that results. In more rigorous attempts to make the centres developmental, designers have included aspects such as self or peer assessment, continuous feedback throughout the centre, and input from participants in writing their final reports or sessions on self–development during the centre. An important aspect of designing successful assessment or development centres is for organisations to consider the true purpose of the centre at the outset, and to be honest about this. If a development centre contains elements of assessment then its developmental benefits may be limited. Participants will not be in a state of mind to learn or develop if they feel that their careers are on the line. In some cases the term "development centre" can be a euphemism when the real purpose, at least in the minds of the assessors, is assessment.

The employment interview: the importance of structure

Part of the problem with interviews is that they are so commonplace. In most organisations it is taken for granted that interviews will be central to any selection process. The interview is part of the social psychological processes associated with the courtship between the applicant and the organisation. Managers are expected to be able to interview, whether or not they have received any specific training, and they expect to have a say in any appointments made within their area of responsibility. Unfortunately, the interview is not necessarily regarded as an assessment method in the same way as some of the other methods covered in this chapter, and does not always undergo the detailed scrutiny that would precede the introduction of a psychometric ability test or personality questionnaire.

The interview has remained popular despite decades of research cataloguing its many inadequacies and the notoriously low validity of employment interviews (Ulrich and Trumbo, 1965; Arvey and Campion, 1982). Among the many problems that have been identified are the following:

1. Different interviewers come to different conclusions (inter-rater reliability is poor).
2. Interviewers make decisions very quickly and based on minimal information.
3. Interviewers form these rapid judgements on the basis of non-verbal signals such as eye-contact, head nodding and facial expressions.

4. Interviewers remember information mainly from the beginning and end of the interview (primacy and recency effects).
5. Verbal skills predominate the interview.
6. Interviewers are unduly affected by a range of factors unrelated to job performance: physical attractiveness, age, gender, whether they like the candidate, and whether the candidate is similar to the interviewer.
7. Interviewers tend to confirm their hypotheses about candidates rather than looking for disconfirmatory evidence.
8. Interviewers tend to be more influenced by negative information than positive information (unfavourable information effect).

Recently however, there have been some more positive views expressed about the interview. Dreher and Maurer (1989) criticise much of the research into interview validity, and conclude that managers may have been correct to ignore the researchers' pessimism about the usefulness of interviews. Wiesner and Cronshaw (1988) used meta-analysis techniques to look at different interview formats. They found that structure has an important effect on interview validity: structured interviews had substantially higher validities than unstructured interviews.

It is likely that the way forward for interviewing is through the development of appropriate structures that managers can be trained to use, and which focus the interview on job-relevant information. Well-researched examples of structured approaches to interviewing include the Patterned Behaviour Description Interview (PBDI), (Janz, 1982; 1989) and the Situational Interview (Latham *et al.*, 1980; Latham, 1989). Both of these methods involve collecting thorough job analysis data by exploring the critical incidents experienced by job holders. They also both use job-related questions which are standardised for all candidates. The PBDI concentrates on asking candidates about examples of experiences they have had that are close to the critical incidents elicited during the job analysis. For example, the candidate might be asked about the last time he or she had to complete some work within a tight time-scale. The responses to the questions are scored on predetermined scales, and further examples elicited to look for patterns of behaviour. In contrast, the Situational Interview uses hypothetical events based on the job analysis. The candidate is asked "What would you do if . . .?" rather than "What did you do when . . .?". The interviewer then compares the answer with a scoring system for evaluating the different possible responses. Both the PBDI and the Situational Interview have been shown to be valid predictors of job performance, achieving better results than traditional interviews. Further research is needed to determine whether the validity of these methods is incremental to the validity provided by ability tests, and whether the methods should be used together or separately, but they do provide grounds for optimism that interviews have a legitimate place in occupational assessment.

Work samples

Work sample tests have a long history in occupational assessment probably because it seems common sense to use assessment methods that simulate job tasks as closely as possible. To support this view, work samples have been shown to be strong predictors of job performance (Asher and Sciarrino, 1974; Schmitt *et al.*, 1984). They are also highly acceptable to candidates because of their perceived closeness to the job, and this may have benefits in helping to reduce adverse impact (Robertson and Kandola, 1982). The classical example of a work sample is to ask applicants for a secretarial position to complete a typing test, and it is in manual tasks such as this that it is possible to design tests that are very close to the actual work that needs to be performed.

This is more difficult with managerial jobs, where the closest method to work sampling is probably the assessment centre. Some organisations have developed further the work-sampling approach by designing the centre around a typical day in the life of the manager. This includes dealing with paperwork (in-tray exercises), meetings (group exercises and one-to-one role plays), presentations, etc., and may also include interruptions such as the telephone ringing in the middle of working on a written report, or the need for candidates to manage their own time in completing tasks, and arranging meetings. This type of assessment centre overlaps with business simulations and games, especially where the consequences of earlier decisions affect events later in the day. The important benefits of realism in the simulation have to be balanced against the need for standardisation and control to ensure that assessments are reliable.

FUTURE DEVELOPMENTS IN OCCUPATIONAL ASSESSMENT

A section on the future developments in almost any field is likely to mention the possibilities resulting from new technology. In occupational assessment the two most likely areas for future developments are new technology and new psychology.

With regard to new technology, the computerisation of assessment is still in its infancy. It is surprising to note that despite the widespread availability of powerful personal computers, much of the testing in organisations is still conducted with pencil and paper administration, and manual scoring. Developments such as multimedia and high-speed video compression should make possible a host of realistic interactive simulation exercises for assessment which can be run on equipment found in every office. This should allow greater use of work sampling outside its traditional application to manual and clerical jobs.

Another area of new technology that has gained a great deal of publicity is the networking of computers. Already millions of computers are linked across the world, and in future the personal computer, with communications capabilities including video, is likely to be common in homes as well as offices. This opens the possibility for remote assessment. Job applicants for example could be tested and even interviewed on-line in their own office, educational institution or home, thus avoiding the need to take time off work to attend the early stages of a selection procedure. This could also make more systematic assessment available to smaller companies who do not have the resources to train staff in the use of psychometric instruments, but who could make use of remote assessment bureaux. Candidates could be assessed on-line at the company's offices during an interview day with the results interpreted remotely. These types of development are possible from a technological perspective, but the pace of change may be slower and will be driven by the preferences of the users of occupational assessment.

New psychology may also lead to changes in assessment in the workplace. As the research and theoretical base expands in other fields, so too will our understanding of people's behaviour, thinking processes, attitudes, emotions and motivations at work. In the immediate future this is likely to come from importing developments from fields such as cognitive, social, and clinical psychology and from changes of emphasis in occupational assessment.

A historical example of importing ideas from other areas of psychology is the use of personality questionnaires, many of which were originally designed for clinical use. The development of specifically work-related questionnaires tended to come later. A more recent example is the application of attribution theory to the workplace. This theory which deals with how people attribute causes to events has generated a large volume of research (e.g. see Hewstone, 1989) and has had considerable impact on social psychological theory and clinical practice, but its possibilities for organisational use have only emerged recently.

Attribution theory has been applied to understanding the interview (Herriot, 1989), but its greatest benefits are likely to come from an understanding of personal attributional styles. Seligman and Schulman (1986) demonstrated that those with an optimistic style were more successful in a sales role than those with a pessimistic style. Optimists, for example, tend to put negative events down to external, temporary and specific causes whereas pessimists tend to put negative events down to personal, stable, and general causes ("It's my fault, it will happen again in the future, and it will affect lots of areas of my life"). More recently human resource professionals have applied attribution theory in a greater variety of contexts under the banner of Attribution Management, including selection and assessment for jobs at many levels, and the diagnosis of development needs. The latter area offers promise for improving performance at the individual, team, or organisational level, as there is evidence that attributional style can be changed (Forsterling, 1985) through training and self development.

Changes in emphasis in occupational assessment are also needed in the future, alongside new approaches. An example might be a move away from assessment of the individual's capability to do the job which is the focus of current tests, questionnaires and simulation exercises to methods which address the person-organisational fit, or what Smith (1994) calls the relational domain, in his theory of the validity of predictors for selection. This might lead, for example, to more use of realistic job previews to give applicants a greater insight into the culture and values of the organisation. Other changes in emphasis could come as a result of greater attention being paid by practitioners to the research comparing the validity of the various assessment methods on offer. There is plenty of scope for assessment to be more focused on the techniques described in this chapter which have a closer relationship to job content and which have been demonstrated to be the most valid predictors of job performance.

Change in occupational assessment is inevitable. It will be interesting to see whether future developments can be influenced by advances in theory and research such as those described in this chapter, or will be driven solely by the commercial and political climate in organisations which determine the priorities of managers.

References

Abbott, D. (1996) Lessons from lessons. In Croll, P. (ed.) *Teachers, Pupils and Primary Schooling*. London: Cassell.

Adler, S. (1987) Towards the more efficient use of assessment center technology in personnel selection. *Journal of Business and Psychology*, **2**, 74–93.

Ahmad, E. (1993) *Report on Preventive Measures being Adopted by the Punjab Boards of Intermediate and Secondary Education for Improvement of their Standard*, Lahore: Board of Intermediate and Secondary Education.

Ahmad, E. (1994) *The role of 'Mafia' in disgracing the examination system*, unpublished manuscript.

Airasian, P. (1988) Measurement driven instruction: a closer look. *Educational Measurement: Issues and Practice*, **7**, pp. 6–11.

Aitkin, M. and Longford, N. (1986) Statistical modelling issues in school effectiveness studies (with discussion). *Journal of Royal Statistical Society*, **A149**, 1–43.

Aldrich, V. C. (1963) *Philosophy of Art*. Englewood Cliffs, NJ: Prentice-Hall.

Alexander, L. (1995) Bob Dole is only half right. *The Wall Street Journal*, June 16, p. A14.

Amano, I. (1990) *Education and Examinations in Modern Japan*. transl. by W. K. Cummings and F. Cummings. Tokyo: University of Tokyo Press.

Andrew, B. J. and Hecht, J. T. (1976) A preliminary investigation of two procedures for setting examination standards. *Educational and Psychological Measurement*, **36**, 45–50.

Angoff, W. H. (1971) Scales, norms and equivalent scores. In: R. L. Thorndike (ed.) *Educational Measurement*, Washington, DC: American Council on Education, pp. 508–600.

Angoff, W. H. (1974). The development of statistical indices for detecting cheaters. *Journal of the American Statistical Association*, **69**, 44–49.

Angoff, W. H. (1988) Proposals for theoretical and applied developments in measurement. *Applied Measurement in Education*, **1**, 215–222.

Anrig, G. R. (1988) ETS replies to Golden Rule on "Golden Rule". *Educ. Meas.: Issues and Practice*, **7**, 20–21.

APU (1985) *A Review of Monitoring in Mathematics*. London: Department of Education and Science.

Ardwoino, J. and Berger, G. (1986) L'evaluation comme interpretation. *l'Evaluation au Pour Voir*. POUR 107.

Arvey, R. D. and Campion J. E. (1982) The employment interview: A summary and review of recent research. *Personnel Psychology*, **35**, 281–322.

Asher, J. J. and Sciarrino, J. A. (1974) Realistic work sample tests: a review. *Personnel Psychology*, **27**, 519–533.

Asian Development Bank and Queensland Education Consortium. (1994) *The Royal Government of Cambodia Education Sector Review*. Manila: Asian Development Bank.

Asimov, N. (1994a) Alice Walker story pulled from state test. *San Francisco Chronicle*, February, pp. A1, A13.

Asimov, N. (1994b) Parents fear new exams part of attack on values. *San Francisco*

Chronicle, 7 March, pp. A1, A13.

Assiter, A. and Shaw, E. (eds) (1993) *Using Records of Achievement in Higher Education.* London: Kogan Page.

Ayer, A. J. (1946) *Language, Truth and Logic* 2nd edn. Harmondsworth: Penguin.

Bachor, D. G. and Anderson, J. O. (1993) *Perspectives on Assessment Practices in the Elementary Classroom.* Victoria, BC: Ministry of Education and Ministry Responsible for Human Rights.

Bagnole, J. W. (1977) *TEFL, Perceptions, and the Arab World.* Washington, DC: American Friends of the Middle East.

Baker, E. L. and O'Neil, H. F. (1991) Policy and validatory prospects for performance-based assessment. Chicago: AERA Conference.

Bambrough, J. R. (1979) *Moral Scepticism and Moral Knowledge.* London: Routledge.

Bandura, A. (1977) *Social Learning Theory.* Englewood Cliffs, NJ: Prentice-Hall.

Bardell, G. S., Forrest, G. M. and Shoesmith, D. J. (1978) *Comparability in GCE: A Review of the Boards' Studies, 1964–1977.* Manchester: Joint Matriculation Board.

Bardell, G., Fearnley, A. and Fowles, D. (1984) *The Contribution of Graded Objectives Schemes in Mathematics and French.* Manchester: Joint Matriculation Board.

Barrick, M. R. and Mount, M. K. (1991) The big five personality dimensions and job performance: A meta-analysis. *Personnel Psychology*, **44**, 1–26.

Beardsley, M. C. (1981) *Aesthetics: Problems in the Philosophy of Criticism.* Indianapolis: Hackett.

Beaton, A. E. and Gonzalez, E. J. (1993) Comparing the NAEP trial state assessment results with the IAEP international results. In: L. Shepard, R. Glaser, and R. Linn (eds), *Setting Performance Standards for Student Achievement: Background Studies.* Stanford, CA: National Academy of Education, pp. 371–398.

Beaton, A. E., and Zwick, R. (1990) *Disentangling the NAEP 1985–1986 Reading Anomaly.* Princeton, NJ: Educational Testing Service.

Bennet, S. M., Desforges, C., Cockburn, A. and Wilkinson, B. (1984) *The Quality of Pupil Experiences.* New York: Lawrence Erlbaum.

Berk, R. A. (1986) A consumers' guide to setting performance standards on criterion-referenced tests. *Review of Educational Research*, **56**, 137–172.

Berk, R. A. (ed.) (1984) *A Guide to Criterion-Referenced Test Construction.* Baltimore: Johns Hopkins University Press.

Bernstein, B. (1977) *Class Codes and Control*, Vol. III. London: Routledge.

Best, D. (1985) *Feeling and Reason in the Arts.* London: Allen and Unwin.

Bethell, G. (1989) *A Report on the Examination Systems of Bangladesh.* Cambridge: University of Cambridge Local Examinations Syndicate.

Bhatty, M. A. (1995) The perils of ad hocism. *Dawn* (Karachi), February 20, p. 11.

Biggs, J. (1995) Assessing for learning: some dimensions underlying new approaches to educational assessment. *Alberta Journal of Educational Research*, **41**, 1–17.

Biggs, J. and Collis, K. F. (1982) *Evaluating the Quality of Learning: The SOLO Taxonomy.* New York: Academic Press.

Billington, R. (1988) *Living Philosophy: An Introduction to Moral Thought.* London: Routledge.

Black, P. (1988) *National Curriculum: Task Group on Assessment and Testing.* London: Department of Education and Science.

Blinkhorn, S. F., and Johnson, C. E. (1990) The insignificance of personality testing. *Nature*, **348**, 671–672.

Bock, R. D. (ed.) (1988) *Multi-level Analysis of Educational Data.* San Diego: Academic Press.

Bock, R. D., Gibbons, R. and Muraki, E. (1988) Full information item factor analysis.

Applied Psychological Measurement, **12**, 261–280.
Bonniol, J. (1991) The mechanisms regulating the learning process of pupils. In: P. Weston (ed.), *Assessment of Pupil Achievement: Motivation and School Success*. Amsterdam: Swets and Zeitlinger.
Booth, T. (1983) Policies towards the integration of mentally handicapped children in education. *Oxford Review of Education*, **9**, 255–75.
Bottani, N. (1994) The OECD international education indicators. *Assessment in Education*, **3**, 333–350.
Bowers, W. J. (1964) *Student Dishonesty and its Control in College*. New York: Columbia University, Bureau of Applied Research.
Boyle, E., Crosland, A. and Kogan, M. (1971) *The Politics of Education: Edward Boyle and Anthony Crosland in Conversation with Maurice Kogan*. Harmondsworth: Penguin.
Boyle, S., Fullerton, J. and Yapp, M. (1993) The rise of the assessment centre: a survey of AC usage in the U.K. *Selection and Development Review*, **9**, 1–4.
Bray, M. (1995) *Community and Parental Contribution to Schooling in Cambodia*. Report to UNICEF (East Asia and Pacific Office).
Brennan R. T. and Lockwood, R. E. (1980) A comparison of the Nedelsky and Angoff cutting score procedures using generalizability theory. *Applied Psychological Measurement*, **4**, 219–240.
Brimer, A., Madaus, G. F., Chapman, B., Kellaghan, T. and Wood, R. (1978) *Sources of Difference in School Achievement*. Windsor: NFER Publishing Company.
Broadfoot, P. (1979) Communication in the classroom: the role of assessment in motivation. *Educational Review*, **31**, 3–10.
Broadfoot, P. (1986) Alternatives to Public Examinations. In: D. L. Nuttall (ed.), *Assessing Educational Achievement*. London: Falmer.
Broadfoot, P. (1994) Diagnostic discourse or dead data. *Paper presented at the BERA Annual Conference, St Anne's College, Oxford.*
Broadfoot, P., Murphy, R. and Torrance, H. (eds.) (1990) *Changing Educational Assessment*. London: Routledge.
Broadfoot. P., Osborn, M., Planel, C., Pollard, A. (1995) Teachers and change: a study of primary school teachers' reactions to policy changes in England and France. Paper presented at the Bi-Annual Conference for the Comparative Educational Society of Europe, Copenhagen June 1994. To be published in Winter-Jensen, T. (ed.), *Proceedings*. University of Copenhagen (forthcoming).
Brown, R. (1988) *Group Processes: Dynamics Within and Between Groups*. Oxford: Blackwell.
Burke, J. W., and Jessup, G. (1990) Disentangling validity from reliability. In: *Open University Reader on Assessment Debates*. Buckingham: Open University Press.
Buros, O. (1938 and subsequent editions) *Mental Measurement Yearbook*, New York: Gryphon.
Burstein, L. (1980) Issues in the aggregation of data. In: D. C. Berliner (ed.), *Review of Research in Education*. Washington, DC: American Educational Research Association.
Burton, R. V. (1976) Honesty and dishonesty. In: T. Lickona (ed.), *Moral Development and Behavior*. New York: Holt, Rinehart and Winston, pp. 173–197.
Busch, J. C. and Jaeger, R. M. (1990) Influence of type of judge, normative information and discussion on standards recommended for the National Teacher Examinations. *Journal of Educational Measurement*, **27**, 145–163.
Buss, W. G., and Novick, M. G. (1980) The detection of cheating in standardized tests: Statistical and legal analysis. *Journal of Law and Education*, **9**, 1–64.
Byford, D. and Mortimore, P. (1978) *Examination Results in the ILEA*. Inner London Education Authority: Research and Statistics Group.

C & G (1994) *Policy and Practice for Schemes, Awards and Certification, Assessment and Quality Assurance.* London: City and Guilds.

C & G (1995) *List of Subjects.* London: City and Guilds.

Callen, A. and Geary, B. (1994) Best practice—putting practice testing to work. *Selection and Development Review*, **10**, 4–7.

Callender, C. (1992) *Will NVQs Work? Evidence from the Construction Industry* (IMS No. 228). University of Sussex: Employment Department/Institute of Manpower Studies.

Camara, W. J. and Brown, D. C. (1995) Educational and employment testing: changing concepts in measurement and policy. *Educational Measurement: Issues and Practice*, **14**, 5–11.

Cannell, J. J. (1988) Nationally normed elementary achievement testing in America's public schools: how all 50 states are above the national average. *Educational Measurement: Issues and Practice*, **7**, 5–9.

Carr-Saunders, A. M. and Wilson, P. A. (1933) *The Professions.* Oxford: Oxford University Press.

Cassells, Sir J. (1994) Learning to succeed, the Bolland Lecture. University of the West of England 12.5.94.

CEDEFOP (1994) Competence: the words, the facts. *European Journal of Vocational Training, 1994* (1).

Christie, T. and Forrest, G. M. (1981) *Defining Public Examination Standards.* London: Schools Council.

CIMA (1995) *Setting CIMA's Examinations* by Jean Elley. London: Chartered Institute of Management Accountants.

Cizek, G. J. (1993) Reconsidering standards and criteria. *Journal of Educational Measurement* **30**, 93–106.

Clare, J. (1994) Independent schools accused of league table cheating. *Daily Telegraph*, March 24, p. 1.

CMEC (1992) *CMEC School Achievement Indicators Program. Bulletin* no. 3, July. Toronto: Council of Ministers of Education.

Coleman, J. S., Campbell, E. Q., Hobson, C. J., McPartland, J., Mood, A. M., Weinfield, F. D. and York, R. L. (1966) *Equality of Educational Opportunity.* Washington DC: US Department of Health, Education and Welfare, US Government Printing Office.

Coleman, J. S., Hoffer, T. and Kilgore, S. (1983) *High School Achievement.* New York: Basic Books.

Coleman, J. S. and Hoffer, T. (1987) *Public and Private High Schools: The Impact of Communities.* New York: Basic Books.

Commission for Evaluation of Examination System and Eradication of Malpractice. (1992) *Report* (Chair M. N. Nur), Lahore, Pakistan.

Conant, J. B. (1963) *The Education of American Teachers.* New York: McGraw-Hill.

Corbett, H. D. and Wilson, B. L. (1990) Unintended and unwelcome: the local impact of state testing. Paper presented to AERA Annual Meeting and in: G. W. Nublett and W. T. Fink (eds.), *Testing, Reform and Rebellion.* USA: Ablex.

Creemers, B. (1992) School effectiveness and effective instruction—the need for a further relationship. In: J. Bashi and Z. Sass (eds.), *School Effectiveness and School Improvement.* Jerusalem: Hebrew University Press.

Cresswell, M. J. (1987a) A more generally useful measure of the weight of examination components. *British Journal of Mathematical and Statistical Psychology*, **40**, 61–79.

Cresswell, M. J. (1987b) Describing examination performance: grade criteria in public examinations. *Educational Studies*, **13**, 247–265.

Cresswell, M. J. (1987c) Grade criteria: some unresolved issues. Invited paper presented at

the H.M.I. Conference *GCSE Grade Criteria*. Solihull, 18–19 March.
Cresswell, M. J. (1991) A multilevel bivariate model. In: R. Prosser, J. Rasbash and H. Goldstein (eds.) *Data Analysis with ML3*. London: Institute of Education.
Cresswell M. J. (1994) Aggregation and awarding methods for National Curriculum assessments in England and Wales: a comparison of approaches proposed for Key Stages 3 and 4. *Assessment in Education*, **1**, 45–61.
Cresswell, M. J. and Gubb, J. (1987) *The Second International Mathematics Study in England and Wales*. Windsor: NFER–Nelson.
Cresswell, M. J. and Houston, J. G. (1991) Assessment of the National Curriculum—some fundamental considerations. *Educational Review*, **43**, 63–78.
Cronbach, L. J. (1990) *Essentials of Psychological Testing*. New York: Harper Collins.
Cronbach, L. J., and Webb, N. (1975) Between class and within class effects in a repeated aptitude × treatment interaction: reanalysis of a study by G. L. Anderson. *Journal of Educational Psychology*, **67**, 717–724.
Crooks, T. (1988) The impact of classroom evaluation on students. *Review of Educational Research*, **58**, 438–481.
Cross, A. (1976) *The Question of Max*. New York: Knopf.
Cross, L. H., Frary, R. B., Kelly, P. P. and Impara, J. C. (1985) Establishing minimum standards for essays: blind versus informed reviews. *Journal of Educational Measurement*, **22**, 137–146.
Cross, L. H., Impara, J. C., Frary, R. B. and Jaeger, R. M. (1984) A comparison of three methods for establishing minimum standards on the National Teacher Examinations. *Journal of Educational Measurement*, **21**, 113–129.
Cross Minutes (1887), *Minutes of Evidence Submitted to the Royal Commission chaired by Lord Cross*, Parliamentary Paper 1887, xxix. London: Her Majesty's Stationery Office.
Crowley-Bainton, T. and Wolf, A. (1994) *Access to Assessment Initiative*. Sheffield: Department of Employment.
Curry, L., Wergin, J. F. and associates (1993) *Educating Professionals: Responding to New Expectations for Competence and Accountability*. San Francisco: Jossey–Bass.
Curtain, R. and Hayton, G. (1995) The use and abuse of a competency standards framework in Australia: a comparative perspective. *Assessment in Education*, **2**, 205–225.
CVCP (1986) *Academic Standards in Universities*. London: Committee of Vice-Chancellors and Principals.
CVCP (1993) *TLTP Project ALTER: Handbook for External Examiners in Higher Education*. London: Committee of Vice Chancellors and Principals.
Daily Times (1995) (Lagos), 23 February.
Davie, R., Butler, N. and Goldstein, H. (1972) *From Birth to Seven*. London: Longman.
Davis, S. F., Grover, C. A., Becker, A. H. and McGregor, L. N. (1993). Academic dishonesty: prevalence, determinants, techniques, and punishments. *Teaching Psychology*, **19**, 16–20.
De Gruijter, D. N. M. (1985) Compromise models for establishing examination standards. *Journal of Educational Measurement*, **22**, 263–269.
De Landsheere, G. (1991) General issues relating to pupil achievement. In: P. Weston (ed.), *Assessment of Pupil Achievement, Motivation and School Success*. Amsterdam: Swets and Zeitlinger
Deal, T. E. and Kennedy, A. A. (1982) *Corporate Cultures*. Reading, MA: Addison-Wesley.
Decharms, R. (1984) Motivation enhancement in educational settings. In: R. Ames and C. Ames (eds.) *Research on Motivation in Education*, Vol. 1, *Student Motivation*. Orlando, FL: Academic Press.
Degenhart, R. E. (1990) *IEA Bibliography*. The Hague: IEA Secretariat.

Department of Education and Science (1988) *Task Group on Assessment and Testing: A Report (The Black Report)*. London: DES.

DES (1982) *Examinations at 16-plus: A Statement of Policy*. London: DES/Welsh Office.

Desforges, C. (1989) *Testing and Assessment*. London: Cassell.

Dexter, C. (1977) *The Silent World of Nicholas Quinn*. London: Macmillan.

Dia, M. (1993) *A Governance Approach to Civil Service Reform in Sub-Saharan Africa*. Washington, DC: World Bank.

Digman, J. M. (1990) Personality structure: emergence of the five-factor model. *Annual Review of Psychology*, **41**, 417–440.

Directorate of Extension Programs for Secondary Education (1960) *Studies in Internal Assessment: Procedures and Promotion Policies*. New Delhi.

Dreher, G. F. and Maurer, S. D. (1989) Assessing the employment interview: deficiencies associated with the existing domain of validity coefficients. In: R. W. Eder and G. R. Ferris (eds.), *The Employment Interview: Theory, Research, and Practice*. Newbury Park, CA: Sage.

Dressel, P. L. (1976) Examinations and evaluations in courses. *Handbook of Academic Evaluation*, pp. 208–222. London: Jossey-Bass.

Dronkers, J. (1993) The precarious balance between general and vocational education in the Netherlands. In: A. Wolf (ed.), *Parity of Esteem: Can Vocational Awards Ever Achieve High Status?* London: ICRA, Institute of Education.

DuBois, P. (ed.) (1947) The classification program. *US Army Air Force Aviation Psychology Program Research Report No. 2*. Washington, DC: US Government Printing Office.

Dunbar, S. B., Koretz, D. M., and Hoover, H. D. (1991) Quality control in the development and use of performance assessments. *Applied Measurement in Education*, **4**, 289–304.

Ebel, R. L. (1972) *Essentials of Educational Measurement*. Englewood Cliffs, NJ: Prentice-Hall.

Eckstein, M. A. and Noah, H. J. (1993) *Secondary School Examinations. International Perspectives on Policies and Practice*. New Haven, CT.: Yale University Press.

Ecob, R. and Goldstein, H. (1983) Instrumental variable methods for the estimation of test score reliability. *Journal of Educational Statistics*, **8**, 223–241.

Ellwein, M. C., Glass, G. V. and Smith, M. L. (1988) Standards of competence: propositions on the nature of testing reforms. *Educational Research*, **17**, 4–9.

Elley, W. B. (1992) *How in the World do Students Read?* The Hague: IEA Secretariat.

Epstein, J. and McPartland, J. (1976) The concept and measurement of the quality of school life. *American Educational Research Journal*. **13**, 15–30.

Equal Opportunities Commission (1982) *Do You Provide Equal Opportunities?* Manchester: EOC.

Erfan, N., Bethell, G. S., and Crighton, J. (1995) *Administration, Logistics, and Conduct of Public Examinations*, SERPP Study 9. Islamabad: British Council.

ETS/IAEP (1989) *A World of Differences*. Princeton, NJ: Educational Testing Service.

Fehrmann, M. L., Woehr, D. J. and Arthur, W. (1991) The Angoff cutoff score method: The impact of frame-of-reference rater training. *Educational and Psychological Measurement*, **51**, 857–872.

Filer, A. (1994) Teacher assessment: social process and social product. *Assessment in Education*, **2**, 23–38.

Finn, C. E. (1991) *We Must Take Charge: Our Schools and Our Future*. New York: Free Press.

Fitz-Gibbon, C. T. (1985) A-level results in comprehensive schools. *Oxford Review of Education*, **11**, 43–58.

Fitz-Gibbon, C. T. (1991). Multilevel modelling in an indicator system. In: S. W. Raudenbush and J. D. Willms (eds.), *Schools, Classrooms and Pupils*. New York:

Academic Press.

FitzGibbon, C. T., and Vincent, L. (1995) *Candidates' Performance in Public Examinations in Mathematics and Science*. London: Schools Curriculum and Assessment Authority.

Fitzpatrick, A. R. (1989) Social influences in standard setting: the effects of social interaction on group judgements. *Review of Educational Research*, **59**, 315–328.

Fletcher, S. (1991) *NVQs Standards and Competence. A Practical Guide for Employers, Managers and Trainers*. London: Kogan Page.

Floud, J. E., Halsey, A. H. and Martin, F. M. (1956) *Social Class and Educational Opportunity*, London: Heinemann.

Fogelin, R. J. (1967) *Evidence and Meaning: Studies in Analytic Philosophy*. London: Routledge.

Forrest, G. M. and Orr, L. (1984) *Grade Characteristics in English and Physics*. Manchester: Joint Matriculation Board.

Forrest, G. M. and Shoesmith, D. (1985) *A Second Review of Comparability Studies*. Manchester: Joint Matriculation Board.

Forsterling, F. (1985) Attributional retraining: a review. *Psychological Bulletin*, **98**, 495–512.

Foucault, M. (1979). *Discipline and Punishment: The Birth of the Prison*. Harmondsworth: Penguin.

Foxman, D., Ruddock, G. and McCallum, I. (1990) *APU Mathematics Monitoring 1984–88 (Phase 2)*. London: Schools Examination and Assessment Council.

Franklyn-Stokes, A. and Newstead, S. E. (1995) Undergraduate cheating: who does what and why? *Studies in Higher Education*, **20**, 159–172.

Freedman, K. (1995) Assessment as therapy. *Assessment in Education*, **2**, 102–107.

Fremer, J. (1989) Testing companies, trends, and policy issues: a current view from the testing industry. In: B. Gifford (ed.), *Testing and the Allocation of Opportunity*, Vol. I, Boston: Kluwer, pp. 61–80.

French, S., Slater, J B., Vassiloglou, M. and Willmott, A S (1987) *Descriptive and Normative Techniques in Examination Assessment*. Oxford: University of Oxford Delegacy of Local Examinations.

Galton, F. (1884) *Hereditary Genius*. New York: Appleton.

Gaugler, B. B., Rosenthal, D. B., Thornton III, G. C. and Bentson, C. (1987) Meta-analysis of assessment center validity. *Journal of Applied Psychology*, **72**, 493–511.

Ghiselli, E. E. (1966) *The Validity of Occupational Aptitude Tests*. New York: Wiley.

Gipps, C. (1990) *Assessment: A Teacher's Guide to the Issues*. London: Hodder and Stoughton.

Gipps, C. (1993) Reliability, validity and manageability in large scale performance assessment. Paper presented at AERA Conference, Atlanta, GA.

Gipps, C. (1994) *Beyond Testing*. Sussex: Falmer Press.

Gipps, C. and Goldstein, H. (1983) *Monitoring Children. An Evaluation of the Assessment of Performance Unit*. London: Heinemann.

Gipps, C. and Murphy, P. (1994) *A Fair Test*? Buckingham: Open University Press.

Gipps, C., Steadman, S., Blackstone, T. and Stierer, B. (1983) *Testing Children: Standardised Testing in Local Education Authorities and Schools*. London: Heinemann

Girot, E. A. (1993) Assessment of competence in clinical practice: a review of the literature. *Nurse Education Today*, **13**, 85–90.

Goals 2000: Educate America Act of 1994, Public Law No. 103–227 (1994).

Goldstein, H. (1979) *The Design and Analysis of Longitudinal Studies*. London: Academic Press.

Goldstein, H. (1980) Dimensionality, bias, independence and measurement scale problems in latent trait test score models. *British Journal of Mathematical and Statistical Psychology*, **33**, 234–246.

Goldstein, H. (1983) Measuring changes in educational attainment over time: problems and possibilities. *Journal of Educational Measurement*, **20**, 369–77.

Goldstein, H. (1984) The methodology of school comparisons, *Oxford Review of Education*, **10**, 69–74.

Goldstein, H. (1986a) Gender bias and test norms in educational selection. *Res. Intell.: BERA Newsletter*, May, pp. 2–4.

Goldstein, H. (1986b) Models for equating test scores and for studying the comparability of public examinations. In: D. L. Nuttall, *Assessing Educational Achievement*. London: Falmer.

Goldstein, H. (1989) *Equity in Testing after Golden Rule*. Institute of Education. ERIC Clearing House.

Goldstein, H. (1993) Assessment and accountability. *Parliamentary Brief*, **2**, 33–34.

Goldstein, H. (1995a) *Multilevel Statistical Models*. London: Edward Arnold; New York, Halstead Press.

Goldstein, H. (1995b) *Interpreting International Comparisons of Student Achievement. Educational Studies and Documents* 63. Paris: UNESCO.

Goldstein, H. and Healy, M. (1995) The graphical presentation of a set of means, *Journal of the Royal Statistical Society*, **A158**, 175–177.

Goldstein, H. and Noss, R. (1990) Against the stream. *Forum*, **33**, 4–6.

Goldstein, H. and Sammons, P. (1995) The influence of secondary and junior schools on sixteen-year examination performance. *International Journal of School Effectiveness* (To appear).

Goldstein, H. and Thomas, S. (1995) Using examination results as indicators of school performance. *Journal of the Royal Statistical Society*, A (To appear).

Goldstein, H. and Wood, R. (1989) Five decades of item response modelling. *British Journal of Mathematical and Statistical Psychology*, **42**, 139–167.

Gonczi, A. (1994) Competency based assessment in the professions in Australia. *Assessment in Education*, **1**, 27–45.

Gonzalez, E. J. and Beaton, A. E. (1994). The determination of cut scores for standards. In *Monitoring the Standards of Education* (eds A. C. Tuijnman and T. N. Postlethwaite), Oxford: Pergamon, pp. 171–190.

Good, F. J. and Cresswell, M. J. (1988) *Grading the GCSE*. London: Secondary Schools Examination Council.

Goodlad, J. (1984) *A Place Called School: Prospects for the Future*, New York: McGraw-Hill.

Gould, S. J. (1981) *The Mismeasure of Man*. New York: W. W. Norton.

Grant, G., Elbow, P., Ewens, T., Gamson, Z., Kohli, W., Neumann, W., Olesen, V. and Riesman, D. (1979) *On Competence: A Critical Analysis of Competence-based Reforms in Higher Education*. San Francisco: Jossey-Bass.

Gray, J. (1981) A competitive edge: examination results and the probable limits of secondary school effectiveness, *Educational Review*, **33**, 25–35.

Gray, J., Jesson, D. and Jones, B. (1986) The search for a fairer way of comparing schools' examination results. *Research Papers in Education*, **1**, 91–122.

Gray, J., Jesson, D. and Sime, N. (1990) Estimating differences in the examination performances of secondary schools in six LEAs: a multi-level approach to school effectiveness. *Oxford Review of Education*, **16**, 137–158.

Gray, J., McPherson, A. and Raffe, D. (1983) *Reconstructions of Secondary Education*. London: Routledge.

Gray, J., Reynolds, D., Fitz-Gibbon, C. and Jesson, D. (eds.) (1995) *Merging Traditions: The Future of School Effectiveness and School Improvement*, London: Cassell.

Greaney, V. and Kellaghan, T. (1995) *Equity Issues in Public Examinations in Developing*

Countries. Washington, DC: World Bank.

Greenbaum, W., Garet, M. S. and Solomon, E. R. (1977) *Measuring Educational Progress: A Study of the National Assessment.* New York: McGraw-Hill.

Griffiths, M. and Davies, C. (1993) Learning to learn: action research from an equal opportunities perspective in a junior school. *British Educational Research Journal,* **19**, 43–58.

Guion, R. M. and Gottier, R. F. (1965) Validity of personality inventories in the selection of employees. *Personnel Psychology,* **18**, 135–164.

Gulliksen, H. (1950) *Theory of Mental Tests.* New York: Wiley.

Gunter, B., Furnham, A. and Drakeley, R. (1993) *Biodata: Biographical Indicators of Business Performance.* London: Routledge.

Hadfield, G. (1980) Sources of variation in what constitutes "good" art examinations at age sixteen plus. University of Manchester, unpublished M.Ed. dissertation.

Haladyna, T., Nolen, S. and Haas, N. (1991) Raising standardized achievement test scores and the origins of test score pollution *Educational Research,* **20**, 2–7.

Hall, E. T. (1977) *Beyond Culture.* Garden City: Anchor Books.

Halsey, A. H., Heath, A. F. and Ridge, J. M. (1980). *Origins and Destinations: Family, Class and Education in Modern Britain.* Oxford: Clarendon Press.

Handy, C. B. (1986) *Understanding Organisations.* Harmondsworth: Penguin.

Haney, W. M. (1993) Cheating and escheating on standardized tests. Paper presented at annual meeting of the American Educational Research Association, Atlanta, GA, April.

Haney, W. and Madaus, G. (1986) *Effects of Standardized Testing and the Future of the National Assessment of Educational Progress.* Chestnut Hill, MA: Boston College Center for the Study of Testing, Evaluation and Educational Policy.

Haney, W. M. and Madaus, G. F. (1989) Searching for alternatives to standardised tests: what, whys and whithers. *Phi Delta Kappan,* **70**, 683–687.

Haney, W. M., Madaus, G. F. and Lyons, R. (1993) *The Fractured Marketplace for Standardized Testing.* Boston: Kluwer.

Hansard's Parliamentary Debates, third series, London.

Harlen W. and Qualter, A. (1991) Issues in SAT development and the practice of teacher assessment. *Cambridge Journal of Education,* **21**, 141–151.

Harris, D. and Bell, C. (1994) *Evaluating and Assessing For Learning.* London: Kogan Page.

Harte, N. (1986) *The University of London 1836–1986.* London: Athlone Press.

Hartshorne, H., and May, M. A. (1928) *Studies in the Nature of Character,* Vol. 1, *Studies in Deceit.* New York: Macmillan.

Hennessy, P. (1992) *Never Again. Britain 1945–51.* London: Cape.

Herriot, P. (1989) Attribution theory and interview decisions. In: R. W. Eder and G. R. Ferris (eds.), *The Employment Interview: Theory, Research, and Practice.* Newbury Park, CA: Sage.

Hess, R. D. and Azuma, H. (1991) Cultural support for schooling contrasts between Japan and the United States. *Educational Researcher,* **20**, 2–9.

Hewstone, M. (1989) *Causal Attribution.* Oxford: Blackwell.

Heywood, J. (1994) *Enterprise Learning and Its Assessment in Higher Education.* Report No. 20. Sheffield: Employment Department.

HMI (1991) *Standards in Education 1989–90.* London: HMSO.

Hoggart, R. (1957) *The Uses of Literacy.* London: Chatto and Windus.

Holland, P. W. and Rubin, D. B. (1982) *Test Equating.* New York: Academic Press.

Holmes, E. (1911) *What Is And What Might Be.* London: Constable.

Houston, J. G. (1980) *Report of the Inter-board Cross-moderation study in English Literature at Ordinary Level: 1975.* Aldershot: Associated Examining Board.

Hu, C. T. (1984) The historical background: examinations and control in pre-modern China. *Comparative Education*, **20**, 7–26.

Hunter, J. E. and Hunter, R. (1984) Validity and utility of alternative predictors of job performance. *Psychological Bulletin*, **96**, 72–89.

Hunter, J. E., Schmidt, F. L. and Hunter, R. (1979) Differential validity of employment tests by race: a comprehensive review and analysis. *Psychological Bulletin*, **86**, 721–735.

Hunter, J.E., Schmidt, F. L. and Jackson, G. B. (1982) *Meta-analysis: Quantitative Methods for Cumulating Research Findings Across Studies*. Beverley Hills: Sage.

Hutchison, D. and Schagen, I. (1994) *How Reliable is National Curriculum Assessment?* Slough: National Foundation for Educational Research.

IAASE (1993) *Report on the Work of the IAASE 1992–1993*. London: Independent Appeals Authority for School Examinations.

Inter exams begin amid rowdy scene (1993). *Dawn* (Karachi), May 23.

Jannarone, R. J. (1986) Conjunctive item response theory kernels. *Psychometrika*, **51**, 357–373.

Janz, J. T. (1982) Initial comparisons of patterned behavior description interviews versus unstructured interviews. *Journal of Applied Psychology*, **67**, 577–580.

Janz, J. T. (1989) The patterned behavior description interview: the best prophet of the future is the past. In: R. W. Eder and G. R. Ferris (eds.), *The Employment Interview: Theory, Research, and Practice*. Newbury Park, CA: Sage.

Jenkins, J. (1950) *The Combat Criterion in Naval Aviation*. National Research Council Committee on Aviation Psychology Report No. 6. Washington DC: Division of Aviation Medicine, Bureau of Medicine and Surgery, United States Navy.

Jesson, D. and Gray, J. (1991) Slants on slopes: using multi-level models to investigate differential school effectiveness and its impact on schools' examination results. *School Effectiveness and School Improvement*, **2**, 230–247.

Jessup, G. (1991) *Outcomes. NVQs and the Emerging Model of Education and Training*. London: Falmer.

Johnson, C. and Blinkhorn, S. (1994). Desperate measures: job performance and personality test validities. *The Psychologist*, **7**, 167–170.

Johnson, S. and Cohen, L. (1983) *Investigating Grade Comparability through Cross-moderation*. London: Schools Council.

Karabel, J. and Halsey, A. H. (1977) *Power and ideology in education*. New York: Oxford University Press.

Kariyawasam, T. (1993) Learning, selection and monitoring: Resolving the roles of assessment in Sri Lanka. Paper presented at Conference on Learning, Assessment and Monitoring, International Centre for Research and Assessment, Institute of Education, University of London, July 1993.

Keeves, J. (1992) *The IEA Study of Science III: Changes in Science Education and Achievement 1970–1984*. Oxford: Pergamon.

Kellaghan, T. (1992) Examination systems in Africa: Between internationalization and indigenization. In: *Examinations: Comparative and International Studies*, M. A. Eckstein and H. J. Noah (eds.), Oxford: Pergamon, pp. 95–104.

Kellaghan, T. and Greaney, V. (1992) *Using Examinations to Improve Education. A Study in Fourteen African Countries*, Washington, DC: World Bank.

Kellaghan, T. and Madaus, G. F. (1995) National curricula in European countries. In: E. Eisner (ed.), *Hidden Consequences of a National Curriculum*. Washington, DC: American Educational Research Association.

Kellett, D., Fletcher, S., Callen, A. and Geary, B. (1994) Fair testing: the case of British Rail. *The Psychologist*, **7**, 26–29.

Kennedy, R. (1965) *Congressional Record*, January 26, p. 513.

Kevles, D. J. (1968) Testing the Army's intelligence: psychologists and the military in World War I. *Journal of American History*, **iv**, 565–81.

Knight, P. and Smith, L. (1989) In search of good practice. *Journal of Curriculum Studies*, **21**, 427–440.

Koch, R. and Reuling, J. (eds) (1993) *Modernisation, Regulation and Responsiveness of the Vocational Training System of the Federal Republic of Germany*. Berlin: OECD/BIBB.

Koerner, J. D. (1963) *Miseducation of American Teachers*. New York: Houghton Mifflin.

Kuehn, P., Stanwyck, D. J. and Holland, C. L. (1990) Attitudes toward "cheating" behaviors in the ESL classroom, *TESOL Quarterly*, **24**, 313–317.

Lapointe, A. E. (1986) Testing in the USA. In: *Assessing Educational Achievement* (ed. D. L. Nuttall) London: The Falmer Press, pp. 114–124.

Lapointe, A. E., Askew, J. M. and Mead, N. A. (1992) *Learning Science*. Princeton, NJ: Educational Testing Service.

Latham, G. P. (1989) The reliability, validity, and practicality of the situational interview. In: R. W. Eder and G. R. Ferris (eds.) *The Employment Interview: Theory, Research, and Practice*. Newbury Park, CA: Sage.

Latham, G. P., Saari, L. M., Pursell, E. D. and Campion, M. A. (1980) The situational interview. *Journal of Applied Psychology*, **69**, 569–573.

Lawton, D. and Gordon, P. (1993) *Dictionary of Education*. London: Hodder & Stoughton.

Leake, J. (1995) Britain beats the world at cheating, *Sunday Times*, April 30, p. 5.

Levy, P. and Goldstein, H. (eds.) (1984) *Tests in Education*. London: Academic Press.

Lewin, K. M., and Lu, H. (1991) Access to university education in the People's Republic of China: a longitudinal analysis of the structure and content of the university entrance examination 1984–88. *International Journal of Educational Development*, **11**, 231–244.

Linn, R. (1993) Educational assessment: expanded expectations and challenges, *Educational Evaluation and Policy Analysis*, **15**, 1–16.

Linn, R., Graue, M. and Sanders, N. (1990) Comparing state and district test results to national norms: the validity of claims that "everyone is above average", *Educational Measurement: Issues and Practice*, **9**, 5–14.

Linn, R. L. (1994) Performance assessment: policy promises and technical measurement standards. *Educational Researcher*, **23**, 4–14.

Linn, R. L. and Drasgow, F. (1987) Implications of the Golden Rule settlement for test construction. *Educational Measurement: Issues and Practice*, **6**, 13–17.

Little, A. (1985) The child's understanding of the causes of academic success and failure: a case study of British schoolchildren. *British Journal of Educational Psychology*, **55**, 11–23.

Livingston, S. A. and Zieky, M. J. (1982) *Passing Scores*. Princeton: Educational Testing Service.

Long, H. A. (1985) Experience of the Scottish Examinations Board in developing a grade-related criteria system of awards. Paper presented at the 11th annual conference of the International Association for Educational Assessment held in Oxford, England.

Lord, F. M. (1980) *Applications of Item Response Theory to Practical Testing Problems*. Hillsdale, NJ: Lawrence Erlbaum.

Lord, F. M. and Novick, M. R. (1968) *Statistical Theories of Mental Test Scores*. Reading, MA: Addison-Wesley.

Macaulay, Lord (1898) *The Works of Lord Macaulay*, 12 vols. London: Longmans Green.

McCullagh, P. and Nelder, J. (1989) *Generalised Linear Models*. London: Chapman and Hall.

McGaghie, W. C. (1991) Professional competence evaluation. *Educational Researcher*, **20**, 3–9.

McGaghie, W. C. (1993) *Evaluating Competence for Professional Practice*. In: L. Curry and J. F. Wergin (eds.), *Educating Professionals: Responding to New Expectations for Competence and Accountability*. San Francisco: Jossey-Bass.

McHugh, G., Fuller, A. and Lobley, D. (1993) *Why Take N VQs? Perceptions of Candidates in the South West.* Lancaster University: Centre for the Study of Education and Training.

Mcmeniman, M. (1989) Motivation to learn. In: P. Langford (ed.), *Educational Psychology: an Australian Perspective.* Cheshire: Longman.

Madaus. G. F. (1994) Do we have a crisis in education? The fashioning and amending of public knowledge and discourse about public schools. Division D Vice-Presidential Address, presented at the Annual Meeting of the American Educational Research Association, New Orleans, April.

Madaus, G. F., Airasian, P. W. and Kellaghan, T. (1980) *School Effectiveness: A Reassessment of the Evidence.* New York: McGraw-Hill.

Madaus, G. F. and Kellaghan, T. (1991) Student examination systems in the European Community: lessons for the United States. In: G. Kulm and S. M. Malcom (eds.), *Science Assessment in the Service of Reform*, Washington, DC: American Association for the Advancement of Science, pp. 189–232.

Madaus, G. F. and Kellaghan, T. (1992) Curriculum evaluation and assessment. In: P. W. Jackson (ed.), *Handbook of Research on Curriculum*, New York: Macmillan, pp. 119–154.

Madaus, G. F. and Kellaghan, T. (1993) The British experience with "authentic" testing. *Phi Delta Kappan*, **74**, 458–469.

Madaus, G. F. and Kellaghan, T. (1994) National curricula in European countries. In: E. Eisner (ed.), *The Hidden Consequences of a National Curriculum.* Washington, DC: American Educational Research Association.

Madaus, G., Kellaghan, T. Rakow, E. and King, D. (1979) The sensitivity of measures of school effectiveness. *Harvard Educational Review*, **49**, 207–230.

Mael, F. A. (1991) A conceptual rationale for the domain and attributes of biodata items. *Personnel Psychology*, **44**, 763–792.

Masters, G. N. and Mislevey, R. J. (1991) *New Views of Student Learning: Implications For Educational Measurement.* Princeton: Educational Testing Service.

Mauritius Examinations Syndicate (1994) Issues on the introduction of school-based assessment. *Proceedings of the 1993 IAEA Conference, Mauritius.*

Mavrommatis, I. (1995) Classroom assessment in Greek schools. Ph.D. thesis, University of Bristol School of Education.

Max Planck Institute for Human Development (1983) *Between Elite and Mass Education: Education in the Federal Republic of Germany.* Albany: State University of New York Press.

Mead, G. H. (1934) *Mind, Self and Society.* Chicago: University of Chicago Press.

Mehrens, W. and Kaminski, J. (1989) Methods for improving standardized test scores: fruitful, fruitless or fraudulent? *Educational Measurement: Issues and Practice*, **8**, 14–22.

Mercer, N. (1991) Accounting for what goes on in classrooms. What have neo-Vygotskian's got to offer? *Education Section Review: Journal of the Education Section of the British Psychological Society*, **15**, 61–67.

Merrett, J. and Merrett, F. (1992) Classroom management for project work: an application of correspondence training. *Educational Studies*, **18**, 3–10.

Millman, J. (1984) Computer-based item generation. In: R. A. Berk (ed.), *A Guide to Criterion-Referenced Test Construction.* Baltimore: Johns Hopkins University Press.

Mills, C. N. (1983) A comparison of three methods of establishing cutoff scores on criterion-referenced tests. *Journal of Educational Measurement*, **20**, 283–292.

Mills, R. C. (1991) *The Role of Affect in influencing State of Mind, Self-Understanding and Intrinsic Motivation.* American Educational Research Association.

Mislevey, R. J., Yamamoto, K. and Anacker, S. *et al.* (1991) Towards a test theory for assessing student understanding. *Research Report.* Princeton: Educational Testing Service.

Mislevy, R. J. (1994) Evidence and inference in educational assessment. *Psychometrika*, **59**, 439–484.

Mitroff, I. I. and Kilman, R. H. (1976) On organization stories: an approach to the design and analysis of organizations through myths and stories. In: R. H. Kilman, L. R. Pondy and D. P Slevin (eds.), *The Management of Organization Design: Strategies and Implementation*. New York: North-Holland.

Miyazaki, I. (1976) *China's Examination Hell*. New York: Weatherhill.

Montgomery, R. J. (1965) *Examinations. An Account of their Evolution as Administrative Devices in England*. London: Longman.

Moore, M. (1993) Indian children cheat to survive. *Washington Post*, April 17, pp. A14, A20.

Morley, J. (1903) *Life of William Ewart Gladstone*, 3 vols. London: Macmillan.

Morrison, H. G., Busch, J. C. and D'Arcy, J. (1994) Setting reliable National Curriculum Standards: a guide to the Angoff procedure. *Assessment in Education*, **1**, 181–199.

Mortimore, P., Sammons, P., Stoll, L., Lewis, D. and Ecob, R. (1988) *School Matters: The Junior Years*. Wells, Somerset: Open Books.

Mount, M. K., Barrick, M. R. and Perkins Strauss, J. (1994) Validity of observer ratings of the big five personality factors. *Journal of Applied Psychology*, **79**, 272–280.

Muchinsky, P. M. (1994) A review of individual assessment methods for personnel selection in North America. *International Journal of Selection and Assessment*, **2**, 118–123.

Mullis, I. (1994) *NAEP 1992 Trends in Academic Progress*. Washington, DC: US Department of Education.

Murphy, J. (ed.) (1990) *The Educational Reform Movement of the 1980s*. Berkeley, CA: McCutchan.

Murphy, L. L., Conoley, J. C. and Impara, J. C. (1994) *Tests in Print IV*. Nebraska: University of Nebraska Press.

Murphy, R. J. L. (1982) Sex differences in objective test performance. *British Journal of Educational Psychology*, **52**, 213–219.

Murphy, P. (1989) Assessment and gender. *NUT Education Review*, **3**, 37–41.

Murphy, P. (1991) Assessment and gender. *Cambridge Journal of Education*, **21**, 203–214.

Murphy, R. and Broadfoot, P. (eds.) (1995) *Effective Assessment and the Improvement of Education*. London: Falmer Press.

Naisbitt, J. (1994) *Global Paradox: The Bigger the World Economy, the More Powerful its Smallest Players*. London: Nicholas Brealey Publishing.

Naqvi, S. H. (1993) Examination malpractices, *News* (Karachi), October 14, p. 6.

National Center for Education Statistics. (1987) *The Condition of Education 1987*. Washington, DC: US Department of Education, Office of Educational Research and Improvement.

National Center for Education Statistics. (1991) *The Condition of Education 1991*. Washington, DC: US Department of Education, Office of Educational Research and Improvement.

National Center for Education Statistics. (1994) *The Condition of Education 1994*. Washington, DC: US Department of Education, Office of Educational Research and Improvement.

National Commission on Education (1993) *Learning to Succeed: the Report of the National Commission on Education*. London: Heinemann.

National Commission on Excellence in Education. (1983) *A Nation at Risk*. Washington, DC: US Government Printing Office.

Naylor, F. D. (1990) Student evaluation and examination anxiety. In: H. J. Walberg and G. D. Haertel (eds.), *The International Encyclopaedia of Educational Evaluation*. New York: Pergamon.

NCVQ (1991) *Guide to National Vocational Qualifications.* London: National Council for Vocational Qualifications.

Nedelsky, L. (1954) Absolute grading standards for objective tests. *Educational and Psychological Measurement*, **14**, 3–19.

Newbould, C. A. and Massey, A. J. (1979) *Comparability Using a Common Element.* Cambridge: University of Cambridge Local Examinations Syndicate.

Norcini, J. J. (1990) Equivalent pass/fail decisions. *Journal of Educational Measurement*, **27**, 59–66.

Norcini, J. J. and Shea, J. A. (1993) Increasing pressures for recertification and relicensure. In: L. Curry and J. F. Wergin (eds.), *Educating Professionals: Responding to New Expectations for Competence and Accountability.* San Francisco: Jossey-Bass.

Noss, R., Goldstein, H. and Hoyles, C. (1989) Graded assessment and learning hierarchies in mathematics. *British Educational Research Journal*, **15**, 109–120.

Nuttall, D. L. and Armitage, P. (1984) *A Feasibility Study of a Moderating Instrument.* London: Business and Technician Education Council.

Nuttall, D. L., Goldstein, H., Prosser, R. and Rasbash, J. (1989) Differential school effectiveness. *International Journal of Educational Research*, **13**, 769–776.

Nuttall, D. L. and Willmott, A. S. (1972) *British Examinations: Techniques of Analysis.* Windsor: NFER Publishing Company.

OERI State Accountability Study Group. (1987) *Measuring Up: Questions and Answers About State Roles in Educational Accountability.* Washington, DC: US Department of Education, Office of Educational Research and Improvement.

Office of Technology Assessment-US Congress. (1987) *State Educational Testing Practices: Background Paper.* Washington, DC: Office of Technology Assessment, Science, Education and Transportation Program.

Office of Technology Assessment - US Congress. (1992) *Testing in American Schools: Asking the Right Questions.* Washington, DC: US Government Printing Office.

Ones, D. S., Mount, M. K., Barrick, M. R. and Hunter, J. E. (1994) Personality and job performance: a critique of the Tett, Jackson, and Rothstein (1991) meta-analysis. *Personnel Psychology*, **47**, 147–156.

Ongom, D. L. (1990) Malpractices in examinations. Paper presented at Seminar on Malpractices in Examinations, Zomba, Malawi, September.

Orr, L. and Forrest, G. M. (1984) *Investigation into the Relationship Between Grades and Assessment Objectives in History and English Examinations.* Manchester: Joint Matriculation Board.

Orr, L. and Nuttall, D. L. (1983) *Determining Standards in the Proposed Single System of Examinations at 16+.* London: Schools Council.

OU (1994) *Open University Support for Disabled Students.* Milton Keynes: Open University Press.

Parsloe, P. and Jackson, S. (1992) Changing the culture of professional education: an example from social work. Paper Given at the Colston Society Research Symposium on Language, Culture and Education, Bristol, April.

People's Republic of China. National Education Commission. (1992) *Interim Penal Regulations for Administering the Admission of Students to Ordinary Institutions of Higher Education.* Beijing. Order No. 18.

Perrenoud, P. (1991) Towards a pragmatic approach to formative evaluation. In: P. Weston (ed.), *Assessment of Pupil Achievement: Motivation and School Success.* Amsterdam: Swets and Zeitlinger.

Philippines Congressional Commission on Education. (1993) *Basic Education*, Vol. 1. *The Educational Ladder, Book Two, Making Education Work.* Quezon City: Congressional Oversight Committee on Education.

Philpott, H. B. (1904) *London At School. The Story of the School Board 1870–1904.* London: Fisher Unwin.

Pido, S. (1994) Malpractices in and security of examinations. Paper presented to officials from Pakistan Federal Government and Examination Authority, Kampala, Uganda, May.

Pole, D. (1961) *Conditons of Rational Inquiry: A Study in the Philosophy of Value.* London: Athlone Press.

Plomp, T. (1992) Conceptualizing a comparative educational research framework. *Prospects*, **22** (3), 278–288.

Pollard, A. (1984) Coping strategies and the multiplication of differentiation in infant classrooms. *British Educational Research Journal*, **10**, 33–48.

Pollard, A. (1992) Teachers' responses to the reshaping of primary education. In: M. Arnot and L. Barton (eds.), *Voicing Concerns: Sociological Perspectives in Contemporary Education Reforms.* Wallingford: Triangle Books.

Pollard, A. (1996) Pupil perspectives on curriculum, pedagogy and assessment. In: P. Croll (ed.), *The Social World of Children's Learning.* London: Cassell (Forthcoming).

Pollard, A., Broadfoot, P., Croll, P., Osborn, M. and Abbott, D. (1994) *Changing English Primary Schools? The Impact of the Education Reform Act at Key Stage One.* London: Cassell.

Postlethwaite, T. N. (1987) *Comparative Education Review, Special Issue*, **31** (1).

Postlethwaite, T. N. and Ross, K. N. (1992) *Effective Schools in Reading: Implications for Educational Planners.* The Hague: IEA Secretariat.

Prais, S. (1991) Vocational qualifications in Britain and Europe: theory and practice. *National Institute Economic Review* (May).

Prost, A. (1992) *Education, société et politiques: une histoire de l'enseignement en France, de 1945 a nos jours.* Paris: Seuil.

Quellmalz, E. S. (1990) Essay examinations. In: H. J. Walberg and G. D. Haertel (eds.), *The International Encyclopedia of Educational Evaluation.* New York: Pergamon.

Quicke, J. and Winter, C. (1993) Teaching and language of learning: towards a metacognitive approach to pupil empowerment. Paper presented at the BERA Annual Conference, Liverpool.

Quinlan, M. (1993) Delta index. Paper given at Inter-Group Research Committee seminar on Interpreting Examination Statistics, London, 23 April.

Raudenbush, S. and Bryk, A. (1986) A hierarchical model for studying school effects, *Sociology of Education*, **59**, 1–17.

Raven, J. (1991) *The Tragic Illusion: Educational Testing.* Oxford: Trillium Press.

Resnick, L. (1991) An examination system for the nation. Unpublished paper. University of Pittsburgh.

Resnick, L. B. and Resnick, D. P. (1992) Assessing the thinking curriculum: new tools for educational reform. In: *Changing Assessments: Alternative Views of Aptitude.* In Gifford, B. and O'Connor, M. (eds.), *Achievement and Instruction*, ed., Kluwer. London: Academic Press.

Reynolds, D. and Cuttance, P. (eds.) (1992) *School Effectiveness: Research, Policy and Practice.* London: Cassell.

Roach, J. (1971) *Public Examinations in England 1850–1900.* Cambridge: Cambridge University Press.

Robertson, I. T. and Kandola, R. S. (1982) Work sample tests: validity, adverse impact and applicant reaction. *Journal of Occupational Psychology*, **55**, 171–183.

Robertson, I. T. and Kinder, A. (1993) Personality and job competences: the criterion-related validity of some personality variables. *Journal of Occupational and Organizational Psychology*, **66**, 225–244.

Robinson, W. (1995) How much fifth class math and science do our prospective primary teachers know? Unpublished paper, Primary Education Program Coordination Office, Peshawar, Pakistan.

Rothstein, H. R., Schmidt, F. L., Erwin, F. W., Owens, W. A. and Sparks, C. P. (1990) Biographical data in employment selection: can validities be made generalizable? *Journal of Applied Psychology*, **75**, 175–184.

Rowley, G. L. (1982) Historical antecedents of the standard-setting debate: an inside account of the minimal-beardedness controversy. *Journal of Educational Measurement*, **19**, 87–95.

Rutter, M., Maugham, B., Mortimore, P., Ouston, J. and Smith, A. (1979) *Fifteen Thousand Hours: Secondary Schools and their Effects on Children.* London: Open Books.

Sadler, D. R. (1985) The origins and functions of evaluative criteria. *Educational Theory*, **35**, 285–297.

Sadler, D. R. (1987) Specifying and promulgating achievement standards. *Oxford Review of Education*, **13**, 191–209.

Sadler, D. R. (1989) Formative assessment and the design of instructional systems. *Instructional Science*, **18**, 119–144.

Saint, A. (1987) *Towards a Social Architecture. The Role of School Building in Post-war England.* New Haven: Yale University Press.

Sammons, P., Hillman, J. and Mortimore, P. (1994) *Key Characteristics of Effective Schools: A Review of School Effectiveness Research.* London: Office for Standards in Education.

Sammons, P., Mortimore, P. and Thomas, S. (1996) Do schools perform consistently across outcomes and areas? In: J. Gray *et al.* eds., *Merging Traditions: The Future of Research on School Effectiveness and School Improvement.* London: Cassell.

Sanders, W. and Horn, S. (1994) The Tennessee Value-Added Assessment System: mixed-model methodology in educational assessment. *Journal of Personnel Evaluation in Education*, **8**, 299–311.

Sanderson, M. (1993) Vocational and liberal education—a historian's view. In: A. Wolf (ed.), *Parity of Esteem: Can Vocational Awards Ever Achieve High Status?* London: ICRA, Institute of Education.

Sanderson, M. (1987) *Educational Opportunity and Social Change in England.* London: Faber.

Saupe, J. L. (1960) An empirical model for the corroboration of suspected cheating on multiple-choice tests. *Educ. Psychol. Meas.*, **20**, 475–489.

SCAA (1994a) *GCE A and AS Code of Practice.* London: School Curriculum and Assessment Authority.

SCAA (1994b) *Value-Added Performance Indicators for Schools*, London: School Curriculum and Assessment Authority.

SCAA (1995) *GCSE Mandatory Code of Practice.* London: School Curriculum and Assessment Authority.

Scarth, J. (1984) Teachers attitudes to examining: a case study. In: P. Broadfoot (ed.), *Selection, Certification and Control.* Lewes: Falmer Press.

Scheerens, J. (1992) *Effective Schooling.* London: Cassell.

Schleicher, A. (1994) International standards for educational comparisons. In: *Monitoring the Standards of Education* (eds. A. C. Tujnman and T. N. Postlethwaite), Oxford: Pergamon, pp. 229–248.

Schmidt, F. L. and Hunter, J. E. (1977) Development of a general solution to the problem of validity generalization. *Journal of Applied Psychology*, **62**, 529–540.

Schmidt, F. L., Ones, D. S. and Hunter, J. E. (1992) Personnel selection. *Annual Review of Psychology*, **43**, 627–670.

Schmit, M. J. and Ryan, A. M. (1993) The Big Five in personnel selection: factor structure in applicant and nonapplicant populations. *Journal of Applied Psychology*, **78**, 966–974.

Schmitt, N., Gooding, R. Z., Noe, R. A. and Kirsch, M. (1984) Metaanalyses of validity studies published between 1964 and 1982 and the investigation of study characteristics. *Personnel Psychology*, **37**, 407–422.

Schmitt, N. and Noe, R. A. (1986) Personnel selection and equal opportunity. In: C. L. Cooper and I. T Robertson (eds.), *International Review of Industrial and Organizational Psychology*. Chichester: Wiley.

Schools Council (1979) *Standards in Public Examinations: Problems and Possibilities. Report from the Schools Council Forum on Comparability*. London: Schools Council.

Schwab, J. J. (1989) Testing and the curriculum *Journal of Curriculum Studies*, **21**, 1–10.

Scott, R. (1991) Reflective self-awareness: a basic motivational process. Paper given at the Annual AERA Conference.

SEC (1984) *The Development of Grade-related Criteria for the General Certificate of Secondary Education—a Briefing Paper for Working Parties*. London: Secondary Examinations Council.

SEC (1985) *Reports of the Grade-related Criteria Working Parties*. London: Secondary Examinations Council.

SEC (1986) Draft grade criteria. *SEC News Number 2*. London: Secondary Examinations Council.

SEC (1987) Grade criteria—progress report. *SEC News Number 6*. London: Secondary Examinations Council.

Seligman, M. E. P. and Schulman, P. (1986) Explanatory style as a predictor of productivity and quitting among life assurance sales agents. *Journal of Personality and Social Psychology*, **50**, 832–838.

Shakleton, V. and Newell, S. (1991) Management selection: a comparison survey of methods used in top British and French companies. *Journal of Occupational Psychology*, **64**, 23–36.

Shackleton, V. and Newell, S. (1994) European management selection methods: a comparison of five countries. *International Journal of Selection and Assessment*, **2**, 91–102.

Sharp, C., Hutchison, D. and Whetton, C. (1994) How do season of birth and length of schooling affect children's attainment at key stage 1? *Educational Research*, **36**, 107–121.

Shavelson, R. J., Baxter, G. P. and Gao, X. (1993) Sampling variability of performance assessments. *Journal of Educational Measurement*, **30**, 215–232.

SHC issues notice to chairman in case challenging enquiry (1995) *News International* (Karachi), February 13.

Shepard, L. (1980) Standard-setting issues and methods. *Applied Psychological Measurement*, **4**, 447–467 .

Shepard, L., Camilli, G. and Averill, M. (1981) Comparison of procedures detecting test item bias with both internal and external ability criteria. *Journal of Educational Statistics*, **6**, 317–75.

Silver, H. (1993) *External Examiners: Changing Roles?* London: Council for National Academic Awards.

Silver, H., Stennett,A. and Williams, R. (1995) *The External Examiner System: Possible Futures*. London: UK Higher Education Quality Council.

Simon, B. (1991) *Education and the Social Order 1940–1990*, London: Lawrence and Wishart.

Simon, B. (1978) *Intelligence, Psychology and Education: A Marxist Critique*. London: Lawrence and Wishart.

Smith, M. (1994) A theory of the validity of predictors in selection. *Journal of Occupational and Organizational Psychology*, **67**, 13–31.

Smith, M. and Abrahamsen, M. (1994) More data on the use of selection methods in the U.K. *Selection and Development Review*, **10**, 6–8.

Smith, M. L. (1991a) Meanings of test preparation. *American Educational Research Journal*, **28**, 521–542.

Smith, M. L. (1991b) Put to the test: the effects of external testing on teachers. *Educational Research*, **20**, 8–11.

Snyder, T. D. (ed.) (1991) *Digest of Education Statistics 1991*. Washington, DC: US Department of Education, Office of Educational Research and Improvement.

Spearman, C. (1904) "General intelligence" objectively determined and measured. *American Journal of Psychology*, **15**, 201–293.

Stake, R. (1991) Impact of changes in assessment policy. In: *Advances in Program Evaluation*, Vol. 1, *Using Assessment Policy to Reform Education*. London: Jai Press.

Stancavage, F. B., Roeber, E. and Borhnstedt, G. H. (1992) A study of the impact of reporting the results of the 1990 Trial State Assessment: First report. In: *Assessing Student Achievement in the States: Background Papers*, Stanford, CA: National Academy of Education, pp. 259–2840.

Stedman, L. C. and Smith, M. (1983) Recent reform proposals for American education. *Contemporary Education Review*, **27**, 85–104.

Steedman, J. (1980) *Progress in Secondary Schools*. London: National Children's Bureau.

Stobart, G., Elwood, J., Hayden, M., White, J. and Mason, K. (1992b) *Differential performance in examinations at 16+: English and Mathematics*. London: University of London Examinations and Assessment Council.

Stobart, G., Elwood, J. and Quinlan, M. (1992a) Gender bias in examinations: how equal are the opportunities? *British Educational Research Journal*, **18**, 261–276.

Sulganik, L. H. (1994) Apples and apples: comparing performance indicators for places with similar demographic characteristics. *Educational Evaluation and Policy Analysis*, **16**, 125–141.

Sutherland, G. (1973) *Policy-Making in Elementary Education 1870–1895*. Oxford: Oxford University Press.

Sutherland, G. (1984) *Ability, Merit and Measurement: Mental Testing and English Education 1880–1940*, Oxford: Clarendon Press.

Sutherland, G. (1990) Education. In: F. M. L. Thompson (ed.), *The Cambridge Social History of Britain 1750–1950*, vol. iii. Cambridge: Cambridge University Press.

Swanson, D. B., Norman, G. R. and Luin, R. L. (1995) Performance-based assessment: lessons from the health professions. *Educational Researcher*, **24**, 5–11, 35.

Taylor, C. (1994) Assessment for measurement or standards: the peril and promise of large-scale assessment reform. *American Education Research Journal*, **31**, 231–262.

Teddlie, C. and Stringfield, S. (1993) *Schools Make a Difference: Lessons Learned from a Ten-Year Study of School Effects*. New York: Teachers College Press.

TES (1995). Issues of 23 and 30 June 1995. London: Times Educational Supplement.

Tett, R. P., Jackson, D. N. and Rothstein, M. (1991) Personality measures as predictors of job performance: a meta-analytic review. *Personnel Psychology*, **44**, 703–742.

Tharp, R. and Gallimore, R. (1989) *Rousing Minds to Life: Teaching, Learning and Schooling in a Social Context*. Cambridge: Cambridge Univerisity Press.

The Chronicle of Higher Education Almanac (1994b) Faculty attitudes and characteristics in 13 countries and Hong Kong, **XLI** (1), 1 September, pp. 45–6.

The Chronicle of Higher Education Almanac (1994a) Attitudes and activities of full-time faculty members, **XLI** (1), 1 September, p. 35.

The Guardian (1990). Issue of 17 February 1990. London.

The Mail on Sunday (1991). Issue of 17 February 1991. London.

Thom, D. (1986) The 1944 Education Act: "the art of the possible". In: Harold L. Smith (ed.), *War and Social Change. British Society in the Second World War*. Manchester: Manchester University Press.

Thomas, S. and Mortimore, P. (1996) A value-added analysis in one LEA. *Research Papers in Education* (in press).

Thomson, G. (1952) *A History of Psychology in Autobiography*, Vol. 4 In: E. G. Boring, H. S. Langfeld, H. Werner and R. M. Yerkes (eds.), Worcester, MA: Clark University Press.

Tittle C. R. and Rowe, A. R. (1973) Moral appeal, sanction, threat and deviance: An experimental test, *Social Problems*, **20**, 488–497.

Top BISE official removed from office (1994). *News International* (Karachi), October 25.

Torrance, H. (1993) Formative assessment: some theoretical problems and empirical questions. *Cambridge Journal of Education*, **23**, 333–444.

Torrance, H. and Pryor, J. (1995) Investigating teacher assessment at Key Stage 1 of the National Curriculum in England and Wales: methodological problems and emerging issues. *Assessment in Education*, **2**, 305–320.

Towler, L. and Broadfoot, P. (1992) Self-assessment in the primary school. *Educational Review*, **44**, 137–152.

Trice, H. M. and Beyer, J. M. (1984) Studying organizational cultures through rites and ceremonials. *Academy of Management Review*, **9**, 653–669.

Tuijnman, A. C. and Postlethwaite, T. N. (eds.) (1994) *Monitoring the Standards of Education*. Oxford: Pergamon.

Tymms, P. B. and Fitz-Gibbon, C. T. (1990) A comparison of exam boards: 'A' levels. *Oxford Review of Education*, **17**, 17–32.

Tymms, P. B. and Vincent, L. (1995) *Comparing Examination Boards and Syllabuses at A-level: Students' Grades, Attitudes and Perceptions of Classroom Processes: Technical Report*. Belfast: Northern Ireland Council for the Curriculum, Examinations and Assessment.

Tzannatos, Z. (1991) Reverse racial discrimination in higher education in Malayasia: Has it reduced inequality and at what cost to the poor? *International Journal of Educational Development*, **11**, 177–192.

Ulrich, L. and Trumbo, D. (1965) The selection interview since 1949. *Psychological Bulletin*, **63**, 100–116.

Umar, J. (1994) Application of the Rasch scaling in Indonesia. *Paper presented at the Annual General Assembly of the IEA, Yogyakarta, Indonesia, 24 August.*

UNESCO (1992) *Prospects*, **22** (3 and 4).

University of British Columbia Reports (1994) March 24, 1994.

University of Cambridge Local Examinations Syndicate (UCLES) (1990) *Educational Assessment in Sri Lanka: A Report on the Education and Examination Systems in Sri Lanka*. Cambridge.

US Bureau of Labour Statistics (1988) *Occupational Outlook Handbook*. Washington, DC: US Government Printing Office.

US Department of Education, National Center for Education Statistics (1991) *Projections of Education Statistics to 2002*. Washington, DC.

US Departments of Education and Labor (1993) *Preamble to Goals 2000: Educate America Act*. Washington, DC.

van den Dool, P. C. (1993) New trends in VET and FET: policy issues for the late nineties in the Netherlands. *Paper presented at* Workshop on New Trends in Training Policy. Geneva: International Labour Organisation.

Walker, D. (1976) *The IEA Six Subject Survey: An Empirical Study of Education in 21 Countries*. New York: Wiley.

Warren Piper, D. J. (1994) *Are Professors Professional?: the Organisation of University Examinations.* London: Jessica Kingsley.

Weiner, B. (1994) Integrating social and personal theories of achievement striving. *Review of Educational Research,* **64**, 557–573.

Well-schooled in cheating (1993) *Economic Report* (Seoul), March, pp. 34–36.

Weston, P. (ed.) (1991) *Assessment of Pupil Achievement: Motivation and School Success.* Amsterdam: Swets and Zeitlinger.

Wiesner, W. H. and Cronshaw, S. F. (1988) The moderating impact of interview format and degree of structure on interview validity. *The Journal of Occupational Psychology,* **61**, 275–290.

Wigdor, A. and Green, B. (1991) *Performance Assessment for the Workplace.* Washington, DC: National Academy Press.

Wiggins, J. S. (1973) *Personality and Prediction: Principles of Personality Assessment.* Reading, MA: Addison-Wesley.

Willmott, A. S. 1980 *Twelve Years of Examinations Research: ETRU, 1965–1977.* London: Schools Council.

Willms, J. D. (1992) *Monitoring School Performance: A Guide for Educators.* London: Falmer Press.

Wilmut, J. (1981) *A Brief Report on Two Factors which Affect Grade Changes in Mark-remark and Weighting Exercises.* Associated Examining Board Research Report RAC/184. Guildford: AEB.

Wilmut, J. and Rose, J. (1989) *The Modular TVEI Scheme in Somerset: Its Concept, Delivery and Administration.* Report to the Training Agency of the Department of Employment, London.

Wilson, H. W. (1932–94) *Education Index.* New York: H. W. Wilson.

Wirt, J. G. (1995) *Performance Assessment Systems: Implications for a National System of Skill Standards,* Vol. II—Technical Report. Washington: National Governors' Association.

Wirtz, W. (1977) *On Further Examination: Report of the Panel on the Scholastic Aptitude Test Score Decline.* New York: College Entrance Examination Board.

Wise, A. E. (1979) *Legislated Learning: The Bureaucratization of the American Classroom.* Berkeley: University of California Press.

Wolf, A. (1993) *Assessment Issues and Problems in a Criterion-based System.* London: Further Education Unit.

Wolf, A. (1994) *Criterion-Referenced Assessment.* Buckingham: Open University Press.

Wolf, A. (1995) *Competence-Based Assessment.* Milton Keynes: Open University Press.

Wolf, A. and Silver, R. (1995) *Broad Skills Study: Final Report.* Sheffield: Employment Department.

Wolf, T. H. (1972) *Alfred Binet.* Chicago:University of Chicago Press.

Wood, R. (1986) Aptitude testing is not an engine for equalising educational opportunity. *British Journal of Educational Studies,* **34**, 26–37.

Wood, R. (1991) *Assessment and Testing: A Survey of Research.* Cambridge: Cambridge University Press.

Woodhouse, G. and Goldstein, H. (1988) Educational performance indicators and LEA league tables. *Oxford Review of Education,* **14**, 301–319.

World Bank (1990) *Education in Sub-Saharan Africa. Updated Statistical Tables,* AFTED Technical Note No. 1. Washington, DC.

World Bank (1994a) *Governance: The World Bank's Experience.* Washington, DC.

World Bank (1994b) *World Development Report 1994: Infrastructure for Development.* Washington, DC.

Young, M. (1958) *The Rise of the Meritocracy 1870–2033.* Harmondsworth: Penguin.

Index

Index compiled by Geofrey C. Jones

DUNN and CLARK • Applied Statistics: Analysis of Variance and Regression, *Second Edition*
ELANDT-JOHNSON and JOHNSON • Survival Models and Data Analysis
EVANS, PEACOCK, and HASTINGS • Statistical Distributions, *Second Edition*
FISHER and VAN BELLE • Biostatistics: A Methodology for the Health Sciences
FLEISS • The Design and Analysis of Clinical Experiments
FLEISS • Statistical Methods for Rates and Proportions, *Second Edition*
FLEMING and HARRINGTON • Counting Processes and Survival Analysis
FLURY • Common Principal Components and Related Multivariate Models
GALLANT • Nonlinear Statistical Models
GLASSERMAN and YAO • Monotone Structure in Discrete-Event Systems
GROSS and HARRIS • Fundamentals of Queuing Theory, *Second Edition*
GROVES • Survey Errors and Survey Costs
GROVES, BIEMER. LYBERG, MASSEY, NICHOLLS, and WAKSBERG • Telephone Survey Methodology
HAHN and MEEKER • Statistical Intervals: A Guide for Practitioners
HAND • Dicrimination and Classification
*HANSON, HURWITZ, and MADOW • Sample Survey Methods and Theory, Volume I: Methods and Applications
*HANSON, HURWITZ, and MADOW • Sample Survey Methods and Theory, Volume II: Theory
HEIBERGER • Computation for the Analysis of Designed Experiments
HELLER • MACSYMA for Statisticians
HINKELMANN and KEMPTHORNE • Design and Analysis of Experiments, Volume I: Introduction to Experimental Design
HOAGLIN, MOSTELLER, and TUKEY • Exploratory Approach to Analysis of Variance
HOAGLIN, MOSTELLER, and TUKEY • Exploring Data Tables, Trends and Shapes
HOAGLIN, MOSTELLER, and TUKEY • Understanding Robust and Exploratory Data Analysis
HOCHBERG and TAMHANE • Multiple Comparison Procedures
HOEL • Elementary Statistics, *Fifth Edition*
HOGG and KLUGMAN • Loss Distributions
HOLLANDER and WOLFE • Nonparametric Statistical Methods
HOSMER and LEMESHOW • Applied Logistic Regression
HØYLAND and RAUSAND •System Reliability Theory: Models and Statistical Methods
HUBERT • Applied Discriminant Analysis
IMAN and CONOVER • Modern Business Statistics
JACKSON • A User's Guide to Principle Components
JOHN • Statistical Methods in Engineering and Quality Assurance
JOHNSON • Multivariate Statistical Simulation
JOHNSON and KOTZ • Distributions in Statistics
Continuous Univariate Distributions--2
Continuous Multivariate Distributions
JOHNSON, KOTZ, and BALAKRISHNAN • Continous Univariate Distributions, Volume I, *Second Edition;* Volume 2, *Second Edition*
JOHNSON, KOTZ, and KEMP • Univariate Discrete Distributions, *Second Edition*
JUDGE, GRIFFITHS, HILL, LÜTKEPOHL, and LEE • The Theory and Practice of Econometrics, *Second Edition*
JUDGE, GRIFFITHS, HILL, LÜTKEPOHL, and LEE • Introduction to the Theory and Practice of Econometrics, *Second Edition*
JURECKOVA • Robust Statistical Procedures: Asymptotics and Interrelations
KADANE • Bayesian Methods and Ethics in a Clinical Trial

*Now available in a lower priced paperback edition in the Wiley Classics Library.

Applied Probability and Statistics (Continued)

KADANE and SCHUM • A Probabilistic Analysis of the Sacco and Vanzetti Evidence
KALBFLEISCH and PRENTICE • The Statistical Analysis of Failure Time Data
KASPRZYK, DUNCAN, KALTON, and SINGH • Panels Surveys
KISH • Statistical Design for Research
KISH • Survey Sampling
LANGE, RYAN, BILLARD, BRILLINGER, CONQUEST, and GREENHOUSE • Case Studies in Biometry
LAWLESS • Statistical Models and Methods for Lifetime Data
LEBART, MORINEAU, and WARWICK • Multivariate Descriptive Statistical Analysis: Correspondence Analysis and Related Techniques for Large Matrices
LEE • Statistical Methods for Survival Data Analysis, *Second Edition*
LEPAGE and BILLARD • Exploring the Limits of Bootstrap
LEVY and LEMESHOW • Sampling of Populations: Methods and Applications
LINHART and ZUCCHINI • Model Selection
LITTLE and RUBIN • Statistical Analysis with Missing Data
MAGNUS and NEUDECKER • Matrix Differential Calculus with Applications in Statistics and Econometrics
MAINDONALD • Statistical Computation
MALLOWS • Design, Data, and Analysis by Some Friends of Cuthbert Daniel
MANN, SCHAFFER, and SINGPURWALLA • Methods for Statistical Analysis of Reliability and Life Data
MASON, GUNST, and HESS • Statistical Design and Analysis of Experiments with Applications to Engineering and Science
McLACHLAN • Discriminant Analysis and Statistical Pattern Recognition
MILLER • Survival Analysis
MONTGOMERY and MYERS • Response Surface Methodology: Process and Product in Optimization Using Designed Experiments
MONTGOMERY and PECK • Introduction to Linear Regression Analysis, *Second Edition*
NELSON • Accelerated Testing, Statistical Models, Test Plans, and Data Analyses
NELSON • Applied Life Data Analysis
OCHI • Applied Probability and Stochastic Processes in Engineering and Physical Sciences
OKABE, BOOTS, and SUGIHARA • Spatial Tesselations: Concepts and Applications of Voronoi Diagrams
OSBORNE • Finite Algorithms in Optimization and Data Analysis
PANKRATZ • Forecasting with Dynamic Regression Models
PANKRATZ • Forecasting with Univariate Box-Jenkins Models: Concepts and Cases
PORT • Theoretical Probability for Applications
PUTERMAN • Markov Decision Processes: Discrete Stochastic Dynamic Programming
RACHEV • Probability Metrics and the Stability of Stochastic Models
RÉNYI • A Diary on Information Theory
RIPLEY • Spatial Statistics
RIPLEY • Stochastic Simulation
ROSS • Introduction to Probability and Statistics for Engineers and Scientists
ROUSSEEUW and LEROY • Robust Regression and Outlier Detection
RUBIN • Multiple Imputation for Nonresponse in Surveys
RYAN • Statistical Methods for Quality Improvement
SCHUSS • Theory and Applications of Stochastic Difference Equations
SCOTT • Multivariate Density Estimation: Theory, Practice, and Visualization
SEARLE • Linear Models
SEARLE • Linear Models for Unbalanced Data
SEARLE • Matrix Algebra Useful for Statistics
SEARLE, CASELLA, and McCULLOCH • Variance Components